Elemental

Elemental

How the Periodic Table Can Now
Explain (Nearly) Everything

Tim James

ABRAMS PRESS, NEW YORK

Cataloging-in-Publication Data is available from the Library of Congress

ISBN: 978-1-4683-1702-2

Printed and bound in the United States
1 3 5 7 9 10 8 6 4 2

Abrams books are available at special discounts when purchased in quantity for
premiums and promotions as well as fundraising or educational use.
Special editions can also be created to specification. For details, contact
specialsales@abramsbooks.com or the address below.

Abrams Press® is a registered trademark of Harry N. Abrams, Inc.

ABRAMS The Art of Books
195 Broadway, New York, NY 10007
abramsbooks.com

Dedicated to
the students of Northgate High School

Contents

A Recipe for Reality

Fourteen billion years ago, our Universe decided to begin. We don't know what came before (if there *was* a before), we just know it started stretching in every direction and has been doing so ever since.

In the first few nanoseconds after the big bang, all of reality was a glowing soup of particles, frothing at temperatures millions of times hotter than the Sun. As everything spread out, however, things cooled, particles stabilized, and the elements were born.

Elements are the building blocks nature uses for cosmic cooking; the purest substances making up everything from beetroot to bicycles. Studying the elements and their uses is what we call chemistry, although sadly that word has come to mean something sinister for many people.

A writer on a popular health website was recently moaning about "chemicals in our food" and what we can do to keep food "chemical free." These scaremongers seem to think that chemicals are toxins created by lunatics in lab coats, but this view is far too narrow. Chemicals aren't just the bubbling liquids you see in test tubes: they are the test tubes themselves.

The clothes you're wearing, the air you're breathing, and the page you're currently reading are all chemicals. If you don't want chemicals in your food then I'm afraid it's too late. Food *is* chemicals.

Suppose you mix two parts of the element hydrogen with one part oxygen. In scientific notation, you'd write that as H_2O, water, the most famous chemical in the world. Chuck in a bit of the element carbon and you get $C_2H_4O_2$—household vinegar. Multiply each of those ingredients by three and you'll get $C_6H_{12}O_6$, more commonly known as sugar.

The only difference between cooking and chemistry is that while a recipe might specify a vegetable, chemistry wants to go deeper and find out what the vegetable itself is made of. There's practically no limit to what you can describe once you know the elements involved. Consider this beast for example:[1]

$$H_{375,000,000}O_{132,000,000}C_{85,700,000}N_{6,430,000}Ca_{1,500,000}P_{1,020,000}S_{206,000}Na_{183,000}$$

$$K_{177,000}Cl_{127,000}Mg_{40,000}Si_{38,600}Fe_{2,680}Zn_{2,110}Cu_{76,114}Mn_{13}F_{13}Cr_7Se_4Mo_3Co_1$$

It looks like something you might find in a barrel of toxic waste but it's the chemical formula for a human being. You have to multiply each number by seven hundred trillion, but those are the correct chemical ratios for one human body. So, if you hear someone say they distrust chemicals, feel free to reassure them. They are a chemical.

Chemistry is not an abstract subject happening in dingy laboratories: it's happening everywhere around us and everywhere within us.

In order to understand chemistry, therefore, we have to understand the periodic table, that hideous thing you probably remember hanging on the wall of your chemistry classroom. Glaring down at you with all its boxes, letters, and numbers, the periodic table can be intimidating. But it's nothing more than an ingredients list, and once you've learned to decode it, the periodic table becomes one of your greatest allies in explaining the Universe.

So, yes, the periodic table is seriously weird and seriously complicated, but so is the rest of nature. That's what makes it worth studying. That's what makes it beautiful.

CHAPTER ONE

Flame Chasers

THE MOST FLAMMABLE SUBSTANCE EVER MADE

Chemistry really began when we mastered our first reaction: setting fire to stuff. The ability to create and control fire helped us to hunt, cook, ward off predators, stay warm in winter, and manufacture primitive tools. Originally, we burned things like wood and fat, but it turns out that most substances are combustible.

Things catch alight because they come into contact with oxygen, one of the most reactive elements out there. The only reason things aren't bursting into flame all the time is that while oxygen is reactive it needs energy to get going. That's why starting a fire also requires something like warmth or friction. Oxygen has to be heated in order to combust.

The most flammable chemical ever made, though, far worse than oxygen, was created in 1930 by two scientists named Otto Ruff and Herbert Krug.[1] Meet chlorine trifluoride.

Made from the elements chlorine and fluorine in a one-to-three ratio, chlorine trifluoride is unique in being able to ignite literally anything it touches, including flame retardants.

A green liquid at room temperature and a colorless gas when warmed, ClF_3 will set fire to glass and sand. It will set fire to asbestos and Kevlar (the material from which firefighters' suits are made). It will even set fire to water itself, spitting out fumes of hydrofluoric acid in the process.[2]

There are very few instances of ClF_3 being used, though, because it has the inconvenient property of setting fire to almost anything with which it comes into contact. It takes a special kind of maniac to think, "Hmm, I'll give that a go."

The most spectacular ClF_3 incident happened on an undisclosed date at a chemical plant in Shreveport, Louisiana. A ton of it was being moved across the factory floor in a sealed cylinder, refrigerated to prevent it reacting with the metal. Unfortunately, the cold temperature made the cylinder brittle and it cracked, spilling the contents everywhere. The ClF_3 instantly set fire to the concrete floor and burned its way through over a meter in depth before extinguishing. The man moving the cylinder was reportedly found blasted through the air 150 meters away, dead from a heart attack. That was refrigerated chlorine trifluoride.[3]

During the 1940s, a few cautious attempts were made to use it as a rocket fuel, but inevitably it kept setting fire to the rockets themselves so the projects were abandoned.

The only people who made a serious attempt to harness its power were the Nazi weapons researchers of Falkenhagen Bunker.[4] The idea was to use it as a flame-thrower fuel, but it set fire to the flame-thrower and anyone carrying it so, again, it was deemed unusable.

Just think about that. Not only will it set fire to water, chlorine trifluoride is so evil even the Nazis didn't mess with it. What makes it so potent?

The answer is that fluorine behaves in a very similar way to oxygen but needs less energy to get started. It's the most reactive element on the periodic table and effectively out-oxygens oxygen at breaking other chemicals down. So, when you combine it with chlorine, the second most reactive element, you get an unholy alliance that starts fires without encouragement.

FIRE FROM WATER

The Greek philosopher Heraclitus was so enamored with fire he declared it to be the purest substance—the basic matter from which reality was made. According to him, everything was somehow made from fire in one form or another. Fire was, in other words, elemental.

It's an understandable assumption to make since fire does appear to possess magical properties. Then again, Heraclitus lived on a diet of nothing but grass and tried to cure himself of dropsy by lying in a cow shed for three days covered in manure . . . after which he was eaten by dogs.[5] So perhaps we don't need to take Heraclitus's views too seriously.

The reason it was so difficult to identify elements in the ancient world was because, unknown to the early philosophers, very few elements occur in their pure state. Most of them are unstable and combine to form element fusions called compounds.

It works a bit like a singles' bar. Each person is unhappy on their own so they link up with others to form stable pairings. At the end of the evening, most individuals have formed compounds leading to greater stability all around. Only a handful of elements like gold, which doesn't mind being single, remain in their native state.

Almost everything we come across in nature is a compound, so while something like table salt may look pure, the game is being rigged. Table salt is actually a compound of sodium and chlorine—the true elements.

You'll never find a lump of sodium in the ground or a cloud of chlorine drifting on the breeze because both are violently reactive. This makes them virtually undetectable, especially if you're working with the crude lab equipment of the first millennium.

There's also the fact that many elements are shockingly rare. Take the element protactinium used in nuclear physics research; the entire global supply comes from a single flake, weighing 125 g owned by the UK Atomic Energy Authority.[6] With the odds stacked against them, Greek philosophers had no chance of getting things right.

It wasn't until the late seventeenth century that a German experimenter named Hennig Brandt proved everyday substances had elements locked inside them and most of the stuff we thought to be pure, wasn't at all.

On an unknown night in 1669, Brandt was boiling vast quantities of urine in his lab (you've got to have a hobby), probably because urine is gold-colored and he was hoping to make a fortune by solidifying it into the precious metal.

After many hours of what must have been unpleasant work, Brandt was finally left with a thick red syrup and a black residue similar to the gunk you get after burning toast. He mixed these two things together and heated the mixture once more. What happened next made no sense.

His mixture of urine syrup and cooking schmutz suddenly formed a waxy solid, which smelled powerfully of garlic and glowed blue-green. Not only that, it was extremely flammable and gave off blinding white light as it burned. He had somehow extracted fire from water.

Brandt named his chemical *phosphorus* from the Greek for light-bringer, and spent the next six years experimenting with it in secret. And it wasn't a fun six years, either. Each 60-g batch of phosphorus required five and a half tons of urine to be boiled.

Eventually, running out of his wife's money, Brandt went public with the discovery and began selling phosphorus to Daniel Kraft, one of the first science popularizers, who took it around Europe giving demonstrations to amazed royals and scientific institutions.[7]

Brandt, however, kept the method of extraction a closely guarded secret. Although how nobody figured it out has always been a puzzle. He must have had one hell of a cover story to explain why he wanted all that urine.

Nowadays we understand exactly what was going on in Brandt's methods. The human body's recommended intake of phosphorus is between 0.5 and 0.8 g a day, but since everything we eat contains it, we tend to consume over twice that amount. All this excess is passed into the urine and Brandt was just boiling everything else away.

His discovery marked a crucial moment for chemistry because the extracted phosphorus was so markedly different from its source. Urine doesn't glow in the dark (sadly) but it obviously contains a chemical that does. It was proof there were chemicals hiding in plain sight. The elements weren't out of reach.

THE MEN WHO PLAYED WITH FIRE

At the beginning of the eighteenth century, the German chemist Georg Stahl, armed with this new knowledge that everyday substances could be made from hidden elements, decided to put forward an explanation for fire.

When metals burn they form colored powders, which were called calxes at the time. Calxes were notoriously difficult to set alight, so Stahl concluded that they were elements, difficult to ignite because their fire had been removed.

According to this hypothesis, anything flammable contained a substance that escaped into air when heated, leaving behind the charred remains. This substance was named phlogiston from the Greek *phlogizein* (to set alight) and Stahl argued that a fire was phlogiston being separated from a calx.[8]

Stahl's fire hypothesis was important because, unlike previous ideas in chemistry, it was testable. If correct, it should be possible to trap phlogiston and combine it with a calx to regenerate the original metal. By putting forward an idea that could be proven wrong, Stahl gave us a genuine scientific hypothesis and, like most scientific hypotheses, it was quickly destroyed.

The first chink in the armor came from the French-British scientist Henry Cavendish. He was a notoriously shy man with a penchant for collecting furniture, beloved by physicists because he helped provide evidence for the force of gravitation. His greatest contribution to chemistry though was a series of experiments involving acid and iron.

The reaction between these two always released an invisible gas, which Cavendish collected. His first thought was that he had successfully got hold of phlogiston until he discovered something odd. The gas was explosive.[9] If fire was the result of phlogiston escaping, how could phlogiston itself be burned? How could phlogiston escape from itself?

Stranger still, when Cavendish's gas (which he called flammable air) exploded, it generated pure water. If you could make water from other things, maybe water wasn't elemental either.

The next mystery came in 1774 from the heretical English clergyman Joseph Priestley. Priestley was experimenting on calx of mercury (the red powdery substance you get when mercury is burned) and directing beams of sunlight at it with a magnifying glass.[10]

He collected the gas given off and found that other things burned very well inside it, better than they did in normal air. Whatever it was, it was clearly good at removing phlogiston. Logically this gas had to be dephlogisticated because it was able to absorb phlogiston, so he called it "dephlogisticated air."

About two hundred years previously, the Polish magician Michał Sędziwój had discovered air to be a mixture of two gases, one of which was "the food of life" and one of which was useless.[11] Could this be related?

Priestley decided to seal some mice in a box with his dephlogisticated gas and they survived without harm. He also discovered, after testing it on himself, that it was actually preferable to regular air and made him feel euphoric to breathe it. Sędziwój's food-of-life gas was apparently the same as his dephlogisticated gas.

Priestley also discovered that plants seemed to breathe the gas out, replenishing a room after a fire had burned. The whole thing was very confusing. Fires generating water, metals generating fire, plants generating air . . . What was going on?

BRINGING ORDER

The answer to all the riddles came in 1775 when Priestley shared his phlogiston results with the French chemist Antoine Lavoisier.

Lavoisier worked for the French government collecting tax contributions but his real passion was science. He had already been experimenting on calxes by the time Priestley's experiments came to his attention[12] and decided it was time to put the phlogiston hypothesis through its paces. If fire was the result of phlogiston leaving a substance, the leftover calx should weigh less.

Priestley had tried taking measurements with his magnifying glass and mercury calx, but precision equipment didn't exist in the eighteenth century. Imagine trying to distinguish a powder weighing 1 g from a powder weighing 1.1 g. Quite the challenge.

Lavoisier decided to scale up Priestley's experiment in order to get a clear result. The difference between 1000 kg and 1100 kg is a difference of 100 kg, which you could see with the naked eye. So, Lavoisier ordered the construction of a nine-foot magnifying glass and blasted a plateful of mercury calx with sunlight.[13]

The results were unmistakable—calxes weighed *more* than the original metal. Everyone had it backward. Fire wasn't the removal of phlogiston: it was something being added from the air itself. Substances like metals and phosphorus were the elements and fire was what happened when they combined with Priestley's gas.

As brilliant as this insight was, Lavoisier wasn't perfect and mistakenly thought Priestley's gas was also responsible for the sour taste of acids. He called it *oxygène* from the Greek *oxys-genes* (sour-maker), which translates into English as oxygen.

The exploding gas Henry Cavendish had isolated was a different element (contained within the acid, not the metal) and, when heated with oxygen, combined to form water. Lavoisier named this gas *hydrogène* from the Greek *hydros-genes* (water-maker), which translates into hydrogen.[14]

This new way of looking at things also explained why you couldn't breathe in a room after a fire had been burning. It wasn't because the fire was giving out a toxic substance: it was because air was partly made from oxygen and fires absorbed it, leaving the other gas behind.

This useless gas was eventually shown to react under extreme conditions and could make niter, one of the key ingredients in gunpowder, so the French statesman Jean-Antoine Chaptal named it *nitregène*—nitrogen.

Science always progresses when a hypothesis is proven wrong and Lavoisier's experiments signed the death warrant on phlogiston. Air was an unreacted mixture of nitrogen and oxygen, water was a fused compound of hydrogen *with* oxygen, and fire was a reaction between oxygen and any available chemical. None of them was an element.

For his efforts, Lavoisier was taken to the guillotine in May 1794. Possibly because he worked as a taxman in pre-revolutionary France (never a good idea), but more likely because he criticized the inferior science of Jean-Paul Marat, who became a leading figure of the revolution. An unlucky end for a great mind, although that's nothing compared to the bad luck of a chemist named Carl Scheele.

THE UNLUCKIEST MAN IN THE HISTORY OF CHEMISTRY

Cavendish, Lavoisier, and Priestley were geniuses of a new science and other people quickly joined the hunt. Everyone wanted the glory of discovering a new element, although agreeing on who makes a discovery isn't always obvious.

Some elements have been around since antiquity so it's impossible to know who originally discovered them. The Old Testament contains passages dating back three thousand years that refer to gold, silver, iron, copper, lead, tin, sulfur (correctly spelled with an f—see Appendix I), and possibly antimony.[15]

Then there are instances of someone predicting an element without actually obtaining a sample. Johan Arfwedson deduced there was an element hidden within petalite rock and named it lithium from the Greek *lithos* (rock), but it wasn't until 1821 that William Brande extracted it.[16]

In order to avoid confusion and settle debates we tend to talk about the first person to isolate an element rather than discover it. Credit goes to the first person who manages to hold a pure sample of an element and recognize it as such. Which brings us to the Swedish chemist Carl Scheele.

In 1772, Scheele successfully made a brown powder, which he named baryte from the Greek *barys*, meaning heavy. He knew there was an element hidden inside (barium) but it was Humphry Davy who isolated it and got the glory.

In 1774, Scheele discovered the gas chlorine (from the Greek *chloros*, meaning green) but didn't realize it was an element. It was again Humphry Davy who made this link in 1808, thus getting the credit.

That same year, Scheele discovered calx of pyrolusite but failed to isolate the elemental manganese inside, achieved a few months later by Johan Gahn.

Then it happened again in 1778 when Scheele identified molybdenum, before it was isolated by Peter Hjelm. And then again in 1781 when he deduced the existence of tungsten but failed to isolate it before Fausto Elhuyar, who got the credit.[17]

Scheele even discovered oxygen in 1771—three years before Priestley—but his manuscript was delayed at the printers and, by the time it was published, Priestley had got his results out.[18]

To commemorate his many contributions to chemistry, the mineral Scheelite was named after him . . . until it was officially renamed calcium tungstate and Scheele was once again nudged out of the history books. If there is a god of chemistry, he apparently hates Carl Scheele.

CHAPTER TWO

Uncuttable

DIAMONDS, PEANUTS, AND CORPSES

In 1812 the German chemist Friedrich Mohs invented a 1 to 10 scale to classify the hardness of minerals. Tooth enamel has a score of 5, for example, while iron ranks as a 4. This means your teeth will technically dent a lump of iron but not the other way around. Although I don't recommend you try it because if you accidentally bite steel (iron with carbon impurity), which has a hardness of around 7.5, you'll regret it.

Diamonds were given a value of 10 because they were the hardest things known at the time. Their claim to the crown was only overthrown in 2003 when a group of researchers from Japan managed to make something even harder—a hyperdiamond.

The most common explanation given for how diamonds form is that coal (fossilized plant) gets compressed underground until it turns hard and transparent. It's what everyone gets told in primary school but it's a complete myth. Diamonds are made in a much more extreme environment.

The same year hyperdiamonds were manufactured, Hollywood birthed its own unbelievable creation: *The Core*, a sci-fi film, which has to be seen to be believed. A few highlights from the movie involve a man hacking the entire global internet from a laptop, sunlight melting the Golden Gate Bridge, and Hilary Swank landing a space shuttle in the San Fernando Valley.

One scene in particular stands out for me. A team of scientists is launched into the Earth's mantle in order to nuke the Earth's core and find themselves dodging diamonds the size of buildings.[1]

What's interesting about this scene is that, while giant diamonds are unlikely, it's otherwise fairly accurate. Diamonds really are made in the Earth's mantle, not in the crust.

A diamond is made solely from carbon and it takes billions of years to grow one. Plants do contain carbon but haven't been around long enough to create the gems we extract from mines today. To fuse carbon into a crystal also takes a staggering amount of pressure and temperature—far more than you could achieve in a planetary crust.

Diamonds are really made a few hundred kilometers into the upper mantle, where pressures are hundreds of thousands times greater than atmospheric pressure and temperatures are comparable to the surface of the Sun. Once they've been made, the crystals are vomited to the surface in volcanic eruptions, which solidify, and we eventually dig them up.

The compressed-plant myth probably arises because we also mine coal and that *is* made from heat-compressed plant, but it forms at wussy temperatures and pressures, inadequate for diamonds.

It is also true that one naturally turns into the other, but it's the opposite of what the myth claims. Diamonds are slightly unstable and will decay into coal over thousands of years. So, the obvious question is: could we reverse the process?

In 2003, Tetsuo Irifune from the Tokyo Institute of Technology decided to try compressing coal into a diamond for real. By using the engineer's equivalent of an extreme pressure cooker, Irifune took a lump of coal-like carbon and subjected it to pressures far in excess of what you'd get in the mantle. The result was a hyperdiamond, a chemical never seen before in nature.[2]

Hyperdiamonds will have a Mohs value greater than 10 but the precise number hasn't been calculated because the original piece of carbon is compressed so much the resulting hyperdiamond is tiny. We're talking a few millionths of a gram.

But we don't have to use coal as our starting material. Dan Frost from the Bavarian Geological Institute in Germany managed to make a diamond by compressing peanut butter,[3] and the Illinois-based company LifeGem can make artificial diamonds by compressing your deceased loved one's ashes. Provided you've got the carbon, it can be crystalized.

The fact that coal, diamond, and hyperdiamond are all made from the same element yet have different properties (we refer to them as "allotropes of carbon") suggests that elements can somehow arrange themselves in different ways.

In order to explain this phenomenon, we're going to have to look closely at the notion of something being diamond-like or "uncuttable." And in ancient Greek the word for uncuttable is one you probably know already: *atom*.

THE MAN WHO PROVED GOD

Imagine holding a grain of sand between your fingertips. It's hard to make out details with the naked eye but logically the grain would have two halves; a left hemisphere and a right one. You could imagine a knife small enough to chop the grain right down the middle, splitting it in two. Then, once you had these half grains, you could repeat the process, slicing to quarter grains and so on.

Theoretically, you could do this forever. No matter how small the grain fragment, you'd always be able to zoom in and divide in half again.

The alternative would make no sense. Imagine chopping a grain up so small that it no longer had a left or right half. A piece so small it didn't have any size and just *was*. For an object like this, the very concept of dividing by two would be meaningless. It would be like trying to divide by two on a calculator and the calculator replying with "Sorry, you have reached the smallest thing, you can't divide anymore." You'd have to be crazy to suggest the existence of a smallest object. Cue Democritus.

Democritus was a philosopher/stand-up comedian living in the fifth century BCE and he took the idea of elemental substances very seriously. He believed everything was made of microscopic uncuttable pieces (atoms) that combined to make the world around us.

Say you've got a packet of M&M's. Rather than eating them in mixed handfuls, every sane human being divides them into piles organized by color and eats them one pile at a time. Don't trust anybody who does otherwise.

This sifting of a mixture into purity is what we're really doing when we break a substance down into its elements; we're grouping the atoms according to type. This would also explain where allotropes come from.

Diamond, coal, and hyperdiamond could all be made from carbon atoms stacked and arranged differently, leading to a variety of properties.

And, as if the atom hypothesis wasn't strange enough, Aristotle later used Democritus's idea to prove the existence of God. Because atoms were constantly in motion, bouncing off each other and flying through the emptiness between, every atom's movement could be back-tracked to a collision with an earlier atom, whose movement could be explained as a collision with an earlier one still. Cause led to effect and every effect had a preceding cause.

If you went back far enough there must have been a first movement that caused everything but had no cause itself. Such a thing (an uncaused cause) would be outside the normal laws of nature while still being able to influence them. God, in other words.[4] Make of that what you will.

LORD OF THE SWAMP

Sadly, along with many other great ideas, Democritus's atomic hypothesis was shelved as the Holy Roman Empire took hold of intellectual Europe. It wasn't until the late 1700s that atoms were given serious attention thanks to the work of an English scientist named John Dalton.

At the age of twelve, most people in England are getting acquainted with being a student in high school. John Dalton was teaching at one. The son of a weaver, Dalton had already taught himself science, mathematics, English, Latin, Greek, and French, and achieved the rank of headmaster by his late teens.[5]

Don't be fooled though. While a fierce academic, Dalton still knew how to have a good time and, like any youngster, spent his free moments collecting samples of swamp gas from local bogs. Surprisingly, he never married.

It was while burning these samples of gas that Dalton learned gases don't react all willy-nilly but combine in specific ratios. Hydrogen and oxygen, for instance, always combine in a two-to-one mix and nothing else. If you have three times as much hydrogen as oxygen, you end up with a third of your hydrogen left at the end. It's as if there's only a limited amount of oxygen "bits" to go around.

Dalton decided the best way of explaining these findings was to assume there were tiny particles making up each elemental gas. Thanks to his proficiency in Greek he was familiar with the work of Democritus and began referring to these particles as atoms.

The idea, however, was not widely accepted. Dalton had a habit of overcomplicating things and the book he published in 1808 to outline his atomic hypothesis was a notoriously difficult read.[6] His ideas were rigorous but his explanations were boring and his chemistry was cumbersome.

Nevertheless, Dalton was greatly respected and was eventually given the privilege of being presented to King William IV. This also led to him committing the biggest faux pas of his career because Dalton was a Quaker and forbidden to wear scarlet clothing, which happened to be the color of robes required for meeting the king. Dalton was color-blind (incidentally, he was the first person to document its existence), and the event's organizers "forgot" to tell him he was wearing robes that would offend his fellow Quakers.[7]

So Dalton went parading around in front of other Quakers in the most outrageous clothing imaginable. The unluckiness of being simultaneously color-blind, a Quaker, and publicly dressed in scarlet is remarkably unfortunate. Somewhere, in a dark corner of purgatory, Carl Scheele is probably cackling to himself.

UNDER PRESSURE

The real watershed for the atomic hypothesis came in 1899 when the French physicist Émile Amagat began experimenting with pressure chambers. Amagat had spent his youth lowering samples of gas into mineshafts to measure how much they got compressed, and by adulthood had designed sophisticated machinery capable of compressing gases to three thousand times atmospheric conditions.

Through these experiments he discovered there was a limit to how far a gas could be squeezed. Once you got to a certain point, the gas fought back and refused to get smaller.[8]

This couldn't be explained with the infinitely-smaller-particles hypothesis. If matter *was* made from infinitely small chunks, then any gas would contain an infinite number of gaps in between them as well. No matter how small you compressed a gas there would always be enough space for the matter to fall into.

The physicist Robert Boyle, son of the Earl of Cork, had conducted experiments on gas pressure and argued that it was possible to compress a gas forever because of this very reason. Amagat's research showed otherwise. A gas had a fixed amount of matter, which meant it probably wasn't made from an infinity of smaller bits.

Combined with Dalton's swamp-gas discoveries, Amagat made the idea of atoms look less like a hypothesis and more like a theory—something that has evidence in its favor. However, there was one big problem or, rather, a very small one. In order to make sense of Amagat's readings you had to accept that atoms were tiny. Unthinkably so.

Imagine looking at planet Earth from space and trying to pick out a single grape on its surface. That is the equivalent of looking at a grape and trying to pick out a single atom on its skin.

If atoms were real they would need to be so small that even waves of visible light would be too big to bounce off them. It wouldn't matter how powerful your microscope was, atoms would be impossible to discern by their very nature.

Scientists are in the business of testing theories once they've been established, but how could you test this one? How could you see the unseeable?

EINSTEIN WAS HERE

Albert Einstein was a legend in his own lifetime. What's more impressive is that he deserved the reputation. Publishing over three hundred scientific papers and essentially inventing the landscape of modern physics, Einstein was the epitome of genius.

It would be foolish to summarize his many achievements in a few paragraphs, so we'll focus on the one most relevant to chemistry: a paper he published on July 18, 1905, in which he made the atomic hypothesis testable rather than speculative.

While working at the Swiss patent office, Einstein stumbled across some research from 1827 by the Scottish botanist Robert Brown. Brown had noticed that grains of pollen floating on water appeared to jiggle in random patterns. Originally, he had assumed the grains were alive but found the same thing happened with sand or dust. The phenomenon was known as Brownian motion and, although unexplained, it was nothing more than a curiosity.

Einstein decided to model the pollen's trajectory through the water and found it could only be explained as the result of bombardment from water particles. To accurately describe how the pollen moved, you had to factor in the friction of pollen against water, which meant you had to accept the existence of "water atoms."

Despite the persistent rumors that he failed math in school, Albert Einstein was a mathematician par excellence and drew up an equation that related water temperature to the pollen grain's likely movement. By introducing an equation with a measurable outcome, Einstein changed the game completely. An idea can be debated but a number cannot, so if you can predict a specific value from your hypothesis you have something to search for directly.

He finished his paper with the phrase, "It is to be hoped that some enquirer may succeed shortly in solving the problem suggested here."[9] As was usually the case with Einstein, his equation was soon tested and confirmed. The zigzagging wasn't random at all, but the result of minor fluctuations in water movement on either side of the grain. Pollen looked like it was undergoing constant collisions because it genuinely was.

In finding this, Einstein did for the atomic hypothesis what Lavoisier did for the elemental one: he provided indisputable, quantitative evidence. You couldn't sensibly discuss elements without atoms anymore, or vice versa. There was no argument to be had. Atoms were real.

CHAPTER THREE

The Machine Gun and the Pudding

THE SMALLEST MOVIE IN HISTORY

In 1989 researchers at IBM pushed the boundaries of marketing by creating a sculpture of their company logo using only thirty-five atoms. Then in 2013 they went even further and created a sixty-second film, *A Boy and His Atom*, by drawing images with atoms and animating them through stop-motion, earning a Guinness World Record for the world's smallest stop-motion film.

By taking hundreds of photographs in different positions and playing them at high speed, the IBM researchers were able to tell the story of an atomic stick-figure who plays with a pet atom. This wasn't an easy thing to do, because as we saw in the previous chapter atoms are too small to be seen.

The trick to their film is that each photograph of the atomic models is not really a photograph. They are images obtained from a scanning tunneling microscope (STM), a device that allows us to peer at distances smaller than visible light can access.

Imagine standing beside a dark hole and dropping a rock over the edge. By timing how long it took to reach the bottom, you could calculate how deep the hole was without being able to see it. STMs work on a similar principle.

The business end of an STM is not a lens but a thin nozzle with tiny particles clinging to the tip. These particles are bound loosely so when you apply an electric current they fall off and land on the surface beneath. As they fall they lose a certain amount of energy, which the STM can measure, calculating how far away the surface is.

By scanning the tip back and forth across an object, any bumps and blips will correspond to a different amount of energy being lost, and the STM can indirectly create a map of what the object must look like.

The filming of *A Boy and His Atom* was carried out by creating a flat sheet of copper and bonding carbon monoxide particles to it in specific positions. As the microscope scanned across the copper, it picked up these carbon monoxides like dots on a Braille picture and created the corresponding image in the computer.[1]

It's a novel idea but how can it be possible? In order to detect the outline of an atom, our STM would need to be dropping particles even smaller than atoms. Where can we find particles that small?

CALL ME "J. J."

At the turn of the twentieth century, the main pursuit of any serious physicist was trying to understand electricity. There were two leading nineteenth-century hypotheses under consideration, each supported by some of the biggest names in science. In one corner was the legendary Hermann von Helmholtz, a staunch believer in particles. He argued that since arcs of electricity cast shadows, for example, it had to be made from matter—electrical atoms.

Leading the opposition was his student, Heinrich Hertz, who preferred to explain things with invisible force fields. Having recently shown that magnetic fields could be used to bend the path of electric current, Hertz argued that electricity also had to be a disturbance in some sort of electric field.[2]

Their disagreement was passionate, although Helmholtz and Hertz remained good friends until the end. Sadly they both died in 1894, shortly before the matter was finally settled by a brilliant British physicist named Joseph John "J. J." Thomson. It was Helmholtz who had been right.

J. J. Thomson was, by anyone's account, a wunderkind of science. He was admitted to the University of Manchester at the age of fourteen and was later appointed to the most prestigious physics post in Britain, taking over from Lord Raleigh as the Cavendish Professor of Experimental Physics at Cambridge University.

The precise details of Thomson's electricity experiments are very mathematical but the premise is straightforward. Fill a small chamber with gas and connect two ends of a circuit to the front and back. At a high voltage it is possible to generate streams of electricity through the gas and, if you place magnets at certain points, you can manipulate their behavior.

By carrying out a variety of studies on this theme, Thomson made several crucial observations. Most important was the fact that electricity moved slowly. Hertz's field hypothesis predicted electricity should move at the speed of light but Thomson's measurements clocked it as practically sluggish by comparison. This meant electricity had mass and was therefore made from particles.

The Irish scientist George Stoney called these particles electrons, from the Greek *electron*, meaning amber (which could be rubbed to create shocks of static electricity), and it caught on. Except electrons were notably different from other particles.

For one thing, the atoms discovered by Dalton and Einstein were two thousand times bigger. In fact, it was possible to shoot a stream of electrons through a plate of solid iron because they could apparently fit through the gaps.

Normal atoms are also happy to approach each other whereas electrons actively repel. This repulsive property was named charge and, in all honesty, it's still a mystery. We can measure its influence and describe the mechanism that causes one electron to repel another, but why electrons have charge is not yet understood.

More pressing for Thomson was the issue of where electrons were coming from. Batteries are composed of regular atoms (bigger and chunkier) so electrons had to be somehow hidden within them. Apparently, atoms weren't the smallest things after all—they contained electrons.

So how come atoms didn't have this property of charge? If the electrons within them were repelling, how were two atoms able to approach each other and even bond?

Thomson concluded that atoms had to contain some additional substance with an anti-charge, canceling the electron charge and giving atoms the appearance of being neutral overall.

Thomson proposed that electrons were nestled within a kind of atomic sponge. Slice away a segment of an atom and you'd see the electrons arranged like plums in a traditional British Christmas pudding. A bit like this:

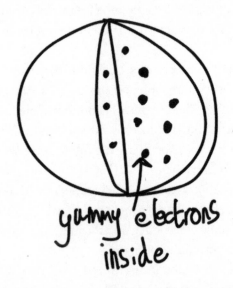

yummy electrons inside

The electrons and dough had opposite and attractive charges, which was why it took so much effort to pull electricity from an atom—you had to rip electrons away from their complementary dough.

Thomson's model of the atom was given the rather catchy name of "The Plum-Pudding Hypothesis."

THE IMPORTANCE OF BEING ERNEST

The name atom had stuck by the time Thomson published his work, which is a shame as it's notoriously misleading. What we call atoms are not uncuttable at all, and neither are they the smallest things. They're just stable structures that prefer not to be pulled apart.

Electrons are the truly uncuttable particles and, as far as Thomson could figure, they were suspended in an oppositely charged dough. But science makes progress by disproving a hypothesis, not by proving it, and the plum-pudding idea was eventually torn to pieces by Thomson's student Ernest Rutherford.

Raised on a New Zealand sheep farm, Rutherford was known for rejecting expensive equipment and carrying out ludicrous experiments because nobody else was doing them. His unorthodox approach earned him the 1908 Nobel Prize in Chemistry, though, so people tended to let him get on with it.

He won the prize for discovering that larger atoms could spit out tiny pieces, which he called alpha particles, that are much heavier than electrons and carry the opposite charge.

Rutherford assumed this happened because atomic dough repelled itself and when the atoms were large there was more chance of self-repulsive instability, leading ultimately to an explosion. The alpha particles he discovered were believed to be bits of atomic dough spat out by the micro-blasts.[3]

Most people would have accepted the Nobel Prize and moved on, but Rutherford was a scientist to the core. He wanted to put his own hypothesis on the chopping block and see if he could disprove it. So he hired the world's best experimentalist, Hans Geiger, and together they developed a method for probing the interior of an atom.

They discovered that alpha particles would produce tiny flashes of light when they hit a piece of zinc sulfide (ZnS), so they spent countless hours sitting in dark rooms firing alpha particles at ZnS, looking for flashes through a lens.

The boredom was unbearable, so Geiger invented an electronic counter that would detect the impacts automatically. His invention was the crackling Geiger counter used in a thousand spy movies ever since.

One morning in 1909, Geiger went to see Rutherford to talk about one of their promising undergraduates, Ernest Marsden. Marsden was only twenty but was gaining a reputation for exceptional lab prowess.

Geiger wanted to give him a new project so Rutherford, in his typical oddball style, came up with something peculiar: "Why not let him see if any alpha particles can be scattered through a large angle in the gold foil experiments?"[4]

The gold-foil experiments had been designed a few years earlier. By taking a piece of radium (a highly alpha-spitting metal) and pointing it at a thin foil you could shoot alpha particles right through the foil. Placing a detector on the other side would let you measure how much the particles were affected by the foil and gave clues about the density of the atomic dough. The best metal was gold because it could be stretched into a leaf only a few atoms thick. The setup looked like this:

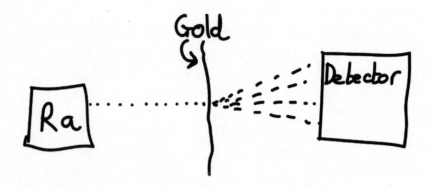

For some reason, Rutherford wanted Geiger and Marsden to put the detector at huge angles to the foil rather than directly on the other side. Geiger must have been puzzled since surely the detector would read nothing but, given Rutherford's reputation (the Nobel Prize medal on his desk probably helped too), he just shrugged and set Marsden to work. The very next day, Rutherford's eccentricity paid off.

The detector began picking up scattered alpha particles even when the detector was moved to the *same side* as the alpha source. This couldn't be explained with the plum-pudding hypothesis because how could an alpha particle get bounced back by dough? It would be like setting up a machine gun to fire at an actual plum pudding and have the bullets bounce back and shoot you in the face. You would expect them to cut right through the pudding and hit the opposing wall, so why are you now in hospital? And what explanation are you going to give the admissions nurse?

Rutherford described it in similar terms: "It was quite the most incredible event that has ever happened to me in my life. It was almost as incredible as if you fired a fifteen-inch shell at a piece of tissue paper and it came back and hit you."[5]

The results were published in February 1910 and, by the following year, Rutherford had run the math. There was only one possible explanation for the result. The atom wasn't a soft sponge all the way through but had hard lumps for the bullets to bounce off. The plum pudding apparently contained nuts.

These nuts were most likely small and clustered in one place within the atom, since only a few bounces were detected for every thousand bullets. It would also make sense for them to have the same charge as alpha particles in order to scatter them when impacted, not to mention holding the electrons in place.

Rutherford called this clump of particles the *nucleus*, from the Latin for nut, and proposed that electrons orbited it like planets around the Sun. Thomson's plum-pudding idea had to be abandoned. It was ingenious, but it had no evidence and in science no evidence means no theory.

YOU WANNA GET NUTS? LET'S GET NUTS!

Did Rutherford have a hunch about the nucleus or was he just fooling around when he suggested moving the detector? Was he trying to think of some task for Marsden and that was the only thing he could come up with at short notice?

Personally, I like to imagine Marsden putting the detector on the wrong side as a sulky middle finger to Rutherford. Here was this great man giving him a stupid task to perform. Oh, you want wide angles? How does the wrong side of the foil sound to you? That wide enough for you, Rutherford?

We'll probably never know but whatever happened in that lab, and whatever went through the minds of the three men, the results have become a part of science lore.

There was still a niggling question that needed answering, though. Rutherford's idea was that the nucleus contained particles with a charge opposite to an electron, but if this was so, why wasn't it ripping itself apart? Particles with the same charge repel each other so the nucleus shouldn't exist at all. The answer was discovered by another of Rutherford's students, James Chadwick, in 1932.

Using a piece of polonium, known to eject alpha particles, Chadwick bombarded a lump of beryllium metal and set up a piece of wax on the other side to cushion any impacts.

Every time there was an emission from the polonium, something inside the beryllium came flying out the other side as if a pool-ball collision was taking place within the nuclei. These ejected particles were obviously heavy but they didn't repel each other, meaning they had to be neutrally charged.

They also had to have some glue-like property that held charged particles together more powerfully than they could repel themselves.

The nut of the atom was apparently made of two types of particle. Neutrons (the neutral ones), which had the glue property, and charged protons (from the Greek word for first), which held the electrons in place. Further research from Niels Bohr, Werner Heisenberg, and Oskar Klein elaborated on Rutherford's findings, and the popular view of the atom was eventually established.

Atoms were like solar systems. Protons and neutrons formed the central nucleus with oppositely charged electrons whizzing around the edge with apparently nothing in between.

If you imagine expanding an atom to the size of a football stadium, an electron would become the size of a dust mote while protons and neutrons would be huddled together in a nucleus, roughly the size of a golf-ball hovering in the center.

The strangest conclusion from this is that most of an atom is empty space. Even something like osmium, the densest element, is apparently 99 percent nothing. As are you.

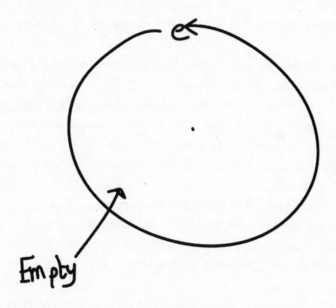

HIDDEN ELEMENTS

In the Superman movie *Man of Steel*, the spaceship that brings Kal-El to Earth is analyzed by chemists and found to be made of elements that don't fit on the periodic table.[6] The periodic table is a list of all known elements so the Kryptonians obviously have different elements on their planet to ours.

The idea of hidden elements is a tantalizing one and it's been thrown around in fiction for decades. In the H. P. Lovecraft story "The Dreams in the Witch House," the protagonist discovers a small statue made of an element that cannot be identified by any scientist.[7] Lovecraft was inspired by a physics lecture he attended in the same year neutrons were discovered, but could such things really exist? Could there be exotic elements tucked away in unknown corners of the Universe?

Not that I want to destroy these fictional stories, but the answer is no. You can't have hidden elements for a straightforward reason: atoms aren't atomic. Yeah, I know, right: what the hell?

The word atom was obviously supposed to mean uncuttable but it's really the electrons, protons, and neutrons that fit this description. The word atom had stuck, however, so we still call them that even though "electron-proton-neutron-superstructures" would be a more accurate term.

The smallest possible atom would logically contain one proton (and one electron since the charges always cancel). This would be element number 1, which turned out to be Cavendish's exploding gas—hydrogen.

The next element would have two protons (plus some neutrons to glue them together). That turned out to be helium. It wouldn't be possible to have element 1.5 in between because there is no such thing as half a proton (see Appendix II).

Once you've got a list of all the elements, you can be sure you haven't missed anything because nature is only able to make atoms in whole numbers. The elements you find on Earth are the same elements you find everywhere in the Universe. Which is where we're going in the next chapter.

So I'm sorry, Superman, your spaceship isn't possible. Interestingly, though, kryptonite is real. The chemical formula for kryptonite is $LiNaSiB_3O_7(OH)F_2$, a mineral which was discovered at a Serbian mine in 2007.[8]

CHAPTER FOUR

Where Do Atoms Come From?

THE COLDEST PLACE IN THE UNIVERSE

The temperature scale we use for our everyday lives was invented in 1742 by Anders Celsius. He took the freezing and boiling temperatures of fresh water, divided the scale into a hundred chunks, and called them "centigrades" from the Latin for hundred steps.

Celsius's original thermometer defined 100°C as freezing and 0°C as boiling, but this was reversed after his death and the scale was renamed the Celsius scale in his honor. The Fahrenheit scale, used more widely in the United States, was invented by Daniel Fahrenheit, who used salted ice and created a scale going up to human body temperature.

Whichever scale you're using, the behavior of particles is the same: as you heat up something, the average speed of its particles increases. Because higher temperatures cause particles to fly around more, this also tells us that when gases get hotter they take up more space. Conversely, if particles become colder they occupy a smaller volume because they move around less. Hotter gas = bigger. Colder gas = smaller.

This simple relationship between temperature and volume is called Charles's law after Jacques Charles, the physicist who discovered it. But obviously this relationship can't go on forever. If you keep cooling things down further and further then the volume shrinks with it, so eventually you should reach a temperature where the volume drops to zero.

Charles's law implies that there's a temperature so cold the particles would take up no space, effectively winking themselves out of existence. This hypothetical temperature is clearly impossible so we call it "absolute zero," calculated to be –273.15°C. It's a temperature so cold you would have to break the laws of physics to reach it.

The coldest place on Earth is usually reported as a point in Antarctica near Dome Argus, which drops to –93.2°C during the winter season.[1]

The emptiness of deep space has an average temperature of –270°C while the Boomerang Nebula stoops to –272°C, one degree above what's physically possible.[2]

But the all-time record for coldest place in the Universe is right here on Earth, at the lab of Martin Zwierlein in Massachusetts, where his team have been able to synthesize a chemical called sodium-potassium, the coldest chemical ever created.

Usually when two atoms bond (see Chapter 8), we attach the suffix ide to whichever element isn't a metal, e.g. iron oxide. A bond between two metal atoms is so rare, however, that we haven't invented a naming system for it, hence the rather unusual-sounding sodium-potassium.

Zwierlein's experiment works by filling a chamber with sodium and potassium atoms in the gas state and heating them to around 7,300°C. By applying a magnetic field across the chamber, the atoms lose their ability to move in as many directions and they begin pairing up (a phenomenon known as Feshbach resonance).

The next step is to zap the gas with two laser beams, one at high energy and one at low energy. When blasted with the high-energy laser, the atoms become stimulated and begin glowing the same color as the beam. Giving out their own light causes them to lose energy, of course, so this is where the second laser comes into play.

Because it is emitting at lower frequency, it serves as a sort of landing platform for the atoms to drop toward. The atoms keep losing energy until they match the frequency of the lower laser, leading to a colossal drop in temperature.

Zwierlein managed to strip the molecules of their heat and plummeted the temperature to five-hundred billionths of a degree above absolute zero, the current world record.[3] But since studying materials at their coldest temperature tells us a lot about how particles behave, we want to go further.

The problem with Zwierlein's experiment is that he performed it on Earth and the gravitational field of our planet pulls slightly on the atoms, causing them to wiggle, thus raising the temperature. The obvious solution is therefore to remove the effects of gravity.

That's the aim of the Cold Atom Laboratory, a version of Zwierlein's experiment due to be performed aboard the International Space Station (ISS). Because the ISS is orbiting the Earth and changing direction constantly, the effects of gravity average to zero. It might be possible to drop atoms not only to billionths of a degree, but to trillionths.

The rules of astrochemistry are clearly very different to those of Earth chemistry and space is where we need to look next in order to understand where elements come from in the first place.

FOR WHAT DO WE KNOW OF THE STARS?

Many centuries ago in the province of Miletus, the great philosopher Thales was ambling through a dark field staring up at the speckled lights that swam across the sky. There were no street lamps in the sixth century BCE so Thales had a perfect view of the Universe with stars uncountable stretching from horizon to horizon.

It was in this moment, as he began to wonder what the stars themselves were made of, that he took a step forward, found nothing but air and toppled into a pit. As he crumpled to the bottom, a Thracian servant girl came dashing up to the edge and giggled hysterically after him, "Maybe you should look to the ground, old man, and not only to the stars!"[4]

I know exactly how he must have felt. I once put my pants on backward while trying to solve an equation in my head. I even did the zip without noticing and only discovered the mistake hours later when I tried to put my hand in my pocket.

Centuries after Thales, the philosopher Aristotle decided that stars were made of an unreachable substance called ether—the holy element of the gods.[5] A nice hypothesis but a completely untestable one since, by definition, gods are beyond the human realm.

If you followed Aristotle's logic, there were unobtainable materials in the Universe and therefore no point in trying to understand what everything was made of.

Unfortunately, his idea caught on and people stopped searching for answers through experimentation and relied on guesswork alone. This trend of trusting opinion over data is why scientific progress died for a millennium and we got stuck in the dark ages. So, nice going, Aristotle.

TWINKLE, TWINKLE

The stranglehold of Aristotle finally began to loosen in 1814 when the German physicist Joseph von Fraunhofer made an important discovery. When you look at a beam of light from a flame, you can split it with a prism and reveal a multitude of colors. It's the same effect that causes rainbows.

What Fraunhofer found was that not every beam of light looks the same when you split it. Different types of fire produce different types of rainbow.

Forty-five years later, Robert Bunsen (of the burner) realized the implications of this discovery. Each element gives off a particular spectrum when burned, like a unique rainbow fingerprint. By studying the light from a fire using Fraunhofer's equipment, you could calculate exactly what atoms are present in the reaction.

This technique, called spectroscopy, allows us to monitor a reaction from a great distance, so if we turn our spectrometers to the stars we should be able to deduce their composition.

The most interesting spectroscopic finding came in 1868 when the French astronomer Pierre Janssen and the British astronomer Norman Lockyer simultaneously observed a completely new elemental signature in the light of our own Sun.[6] It didn't match any of the known elements on Earth so Lockyer named it helium from the Greek *helios*, meaning Sun. Twenty-seven years later, William Ramsay extracted it from terrestrial rocks, making it the only element to be discovered in space before it was isolated on Earth.[7]

The next breakthrough happened in 1925 when the American astronomer Cecilia Payne-Gaposchkin successfully calculated how much of each element was present in a typical star.

Payne-Gaposchkin studied astrophysics at Harvard under Harlow Shapley, one of the only astronomers in the world to let women take the subject, and wrote her PhD thesis on the star classes identified by another astronomer, Annie Jump Cannon (possibly the greatest name in science).

Cannon was completing her nine-volume catalog of every known star when Payne-Gaposchkin began perusing the data. Being well versed in the new science of quantum mechanics (which most astronomers weren't), Payne-Gaposchkin showed that the amounts of each element in stars were vastly different to the amounts found on Earth. Stars weren't just hot planets, as was suggested by the world's leading astronomer Henry Norris Russell, they were something else entirely.[8]

On Earth, the most abundant elements are oxygen, silicon, aluminum, and iron but stars are made almost entirely from hydrogen and helium. The astronomers Otto Struve and Velta Zebergs described Payne-Gaposchkin's research as "undoubtedly the most brilliant PhD thesis ever written in astronomy"[9] but her work was largely dismissed (three guesses why).

Henry Norris Russell even advised her not to publish the results because it would make her a laughing stock but, to his credit, changed his mind when he repeated her methods and found she had been right all along.

The Universe, it turns out, is made almost entirely from hydrogen and helium. The other elements from which we derive all the planets are merely trace impurities. This humbling realization prompted the astronomer Lewis Fry Richardson (or possibly George Gamow, the origin is unclear) to write the following poem in tribute to the discovery:

Twinkle, twinkle little star,
I don't wonder what you are,
For by spectroscopic ken,
I know that you're hydrogen.

STARS UNCOUNTABLE

If you look up on a clear night sky, out in the countryside where there's no light pollution, you can see a pale ribbon of light stretching from horizon to horizon. The ancient Greeks thought it was breast milk from the goddess Hera and called it *galaxias kyklos*, the milky circle.

Today, we know this glowing stream is made from suns. So many suns it becomes impossible to count them as individual points of light so it blurs into a beautiful haze.

The night is filled with what we usually think of as starlight but in a very real sense what we're talking about is sunlight. Our Sun, the source of all our energy, is only one among billions of others orbiting the super-massive black hole Sagittarius A*.

If you were to somehow see our galaxy from the outside, you wouldn't even know our sun was there amid the glow. It would be like looking at a cloud and trying to pick out a single drop of water.

There are somewhere between one and four hundred billion suns in the Milky Way but it's hard to know for sure because we've never been outside it to take a picture. And our galaxy isn't special either. In 964, the Persian astronomer Abd al-Rahman al-Sufi saw what looked like a cloud sitting inside the Andromeda constellation. Little did he realize he had just discovered our nearest galactic neighbor, confirmed in 1923 by the astronomer Edwin Hubble. It sits about twenty quintillion kilometers away from us and contains around a trillion stars.

The telescope that bears Hubble's name, quietly orbiting our planet 547 km above ground, has probed even further than Andromeda and revealed over 170 billion other galaxies in our local region of space.

If someone were to ask how many stars there are in the Universe the answer would sound comical. Even the lowest estimate puts the number of stars at around ten quadrillion in our local region of space alone.

The only kinds of people who talk in numbers that big are preschool children who have no idea how ridiculous the numbers sound and scientists who know *exactly* how ridiculous the numbers sound.

HOW DO STARS FORM?

The usual answer to this question never does justice to the truth. People are usually told that suns are either fires or balls of burning gas. Both views are tragically inadequate. The closest we've ever come to manufacturing the fabric of a star here on Earth was on October 30, 1961. That was when the human race stood in awe and terror as it detonated the Tsar Bomba on the Russian island of Severny.

It is the most powerful nuclear explosion ever achieved to date, with a blast radius of around 35 km. To put things in perspective, our

Sun is equivalent to roughly two billion Tsar Bombas detonating in unison, every second. In one instant, the Sun casually generates over a million times the amount of energy our entire species consumes in a year.

Its light provides the energy needed for our crops to grow, its warmth is what makes water evaporate, giving us rain, and its gravitational pull is what stops us drifting into the cold emptiness of space. It's no exaggeration to say that the Sun maintains the entire human species and permits it to live. And it goes far deeper than that.

To understand what's really going on we'll need to consider the effects of the all-pervading force of gravity, which is usually ignored in chemistry.

All matter in the Universe has a gravity field to it, which means everything is pulling on everything else. We don't feel it but our bodies are loosely gravitating to the objects in the room around us and they are being drawn back toward us in return.

The reason you don't notice this effect is because gravity is a very weak force (you need an entire planet's worth to hold things in place), but while gravity might be weak it is infinite and has been around since the beginning.

Within the first half-second after the big-bang expansion started, the earliest particles called photons and neutrinos (Appendix II again) began colliding, forming the protons, neutrons, and electrons we already know about. A few hundred seconds after that, the protons and neutrons joined up, creating hydrogen and helium nuclei with a tiny bit of lithium and beryllium thrown in (elements 3 and 4). Then, for the next 380,000 years, nothing happened.[10]

During this time the Universe was a buffet of free-floating nuclei and electrons. You wouldn't have been able to see anything in front of your face because there was light in all directions and all of reality would have looked like a milky fog.

Then, after about 1.6 million years, the temperature dropped to a breezy thousand degrees and electrons got snagged by nuclei, forming clouds of hydrogen and helium atoms. The Universe finally became see-through and gravity began exerting its influence.

As the hydrogen/helium clouds started collapsing under their own weight, their gravity fields became more concentrated, pulling more and more atoms into the mix. Over millions of years, these clouds condensed into swirling knots across the Universe, getting hotter and hotter until they whipped themselves into such a frenzy that the nuclei of the atoms began to fuse.

Gravity pulled things inward while the heat from fusion at the core pushed outward. When a truce was finally reached between these forces, the result was a stable sphere of nuclear explosion. The very first sun.

The core of a sun like ours reaches a temperature of about 16 million°C, hot enough to vibrate hydrogen and helium atoms into each other and mash them into heavier elements like oxygen and carbon. Bigger and fiercer stars can go even further, burning carbon atoms into magnesium and then fusing all the way up to iron (element 26). This is how the light elements are made.

TIME TO DIE

In about four billion years, the hydrogen in our Sun will be depleted and things will start to cool. The thermal pressure from within will no longer be hot enough to support its shape and gravity will dominate, causing everything to contract.

This will temporarily increase the core pressure, giving it a momentary second wind of heat and inflating the envelope of gas around the outside of the Sun, making it significantly bigger. At this point the Sun's radius will stretch out to encompass the Earth, burning our beautiful planet to cinders.

As we've already said, though, our Sun is peanuts compared to what else is out there. When big suns reach the end of their lives, something very different happens. A super-giant star will keep burning until its entire core has been converted to iron and, once again, the heat can't support the outer layers and gravitational collapse occurs. But this time we've got a bigger star and more gravity so the contraction happens within seconds. The iron core is too dense to be compressed so when the outer layer shrinks, it bounces off the core and the shockwave causes a catastrophic explosion, which rips the whole thing to pieces.

We call it a supernova and it's during these violent star-plosions that iron atoms get fused together, generating elements all the way up to ninety-two. The star's body has been shredded from the inside out and the newly formed heavy elements are scattered into the dust of space.

And then the whole process repeats. Clouds form, gravity makes them clump, and suns are born, except now we have new atoms in the mixture. The clouds are no longer just hydrogen and helium but colorful mixtures of heavier elements too.

As this second generation of stars is weaved from the corpses of supernovae, the heavier elements get sucked into the star's rotating gravity field. Some of this material gets pulled into the furnace but a lot of it forms a ring, encircling the sun like a moat around a castle.

Clusters of metal and rock gather in the eddies of this current and eventually congeal into planets. Each planet in a solar system is made from atoms that began life inside an ancient star, blown to pieces by the colossal horror of a supernova.

This is not idle speculation either. Thanks to spectroscopy, we have witnessed all of these events happening. The Universe truly is in a cycle of stellar reincarnation with planets and their inhabitants being generated as by-products.

CHILDREN OF STARDUST

There are stories from many cultures about how we are drawn from the dust of the Earth and that we are at one with nature. What science gives us is something far grander: the reassurance that these are not fairy tales.

The first nine months of your life involved your mother building you out of the food she ate, but the atoms in that food came from the Earth and the Earth is made from the remnants of long-dead suns. With the exception of hydrogen, all the atoms in your body started their lives in the heart of a sun, which means you are, as Carl Sagan once observed, made from star stuff.

The stars you see at night are not transcendent objects made from ether as Aristotle believed: they are made from the same material as you. They are your distant relatives and when you die you will return to them. As our planet reaches its fiery demise, your atoms will get spread across the Universe and you will become part of another planet, perhaps even another living being. Maybe the ancient humans who worshipped the stars chose their gods wisely.

CHAPTER FIVE

Block by Block

RECORD-BREAKING FLAVOR

Classifying chemicals according to their properties has been a goal for thousands of years. Today, we use sophisticated equipment but an astonishing amount can be gleaned using our senses.

The human tongue is coated with receptors coming in at least five varieties: sour, bitter, salty, sweet, and umami (sometimes called savory). If the right-shaped chemical docks with a sweet-receptor for instance, a signal is sent to the brain and the food is perceived as sweet. Smell receptors work in the same way, except there are thousands of potential shapes, allowing us to distinguish thousands of fragrances.

The food in our mouth is sensed by the tongue and nose simultaneously. This combination of smell and taste is what gives each food a "flavor." That is, with the exception of spicy foods. They work by accident.

As well as taste, your mouth also needs to monitor the temperature so you don't consume things that are too hot. The heat sensors in your body have names like "TRPV1 receptors" and there are plenty on the tongue and in the gut. Certain chemicals are coincidentally shaped in such a way that they trigger the heat sensors and tell your brain the area is hot, even though the rest of your mouth is cold. The resulting confusion is what we perceive as "spiciness."

In 1912, the American scientist Wilbur Scoville devised a test to measure the spiciness of food mathematically and we still use it today. The spicy chemical is dissolved in water repeatedly until it can no longer be tasted by a panel of volunteers. The number of dilutions required to make the taste imperceptible is then expressed as a Scoville Heat Unit or SHU.

Since the tongue is good at tasting even trace amounts of a chemical, SHU values are typically enormous. The oil from a jalapeño pepper is undetectable after about 8,000 dilutions so jalapeños are given an SHU of 8,000, while something like Tabasco sauce scores closer to 50,000.[1]

The world's spiciest pepper at the time of writing is the Dragon's Breath chili, bred into existence by Welsh spice-master Mike Smith and possessing an SHU of over 2.4 million.[2] That's basically the same as pepper spray. This chili is so hot it would trigger anaphylactic shock if you ate it, but that's nothing compared to the world's spiciest chemical: resiniferatoxin.

Produced in the latex of *Euphorbia resinifera* plants (also called resin spurges), nobody has ever carried out taste trials with resiniferatoxin because it is acutely toxic and causes severe burns to the skin, meaning we have to calculate its SHU indirectly.

A study carried out by Arpad Szallasi in 1989 (on rats) found that resiniferatoxin was one thousand to ten thousand times better at binding to TRPV1 receptors than the chemical in chili peppers.[3] Since we know chili peppers have an SHU of around 16 million, resiniferatoxin is going to score somewhere in the region of 16–160 billion SHU. That's spicy enough to kill you.

There are many other chemicals that have record-breaking effects on our senses too. The sweetest chemical, so sickly it induces vomiting, is called lugduname, 230,000 times sweeter than table sugar.[4]

The darkest chemical, so black you can't even see a torch shining on it, is called vantablack.[5]

And the worst-smelling chemical is a tie between propanthione and methanethiol, substances that have caused mass unconsciousness, spontaneous vomiting, and even death from smelling them at a distance.[6]

THE ELEMENT WAR

The first attempt at properly identifying elements was done by none other than Pythagoras himself, although it was a little bit weird. Most people know Pythagoras from the square-on-the-hypotenuse law they learned in school. What's not usually mentioned is that Pythagoras was also probably the world's first cult leader.

Not a great deal is known about the Pythagorean order because revealing their secrets would get you exiled, but we know they were forbidden from touching white chickens or eating beans.[7] Pythagoras was murdered because an angry mob chased him to the edge of a bean field and, rather than entering it, he turned toward the crowd and chose a fatal beating.[8]

The only other thing we know about the Pythagoreans is that they considered numbers to be elements. Pythagoras and his cult worshipped numerical order, believing math to be the true face of reality. Their table of elements was simply a list of numbers going from one upward. Okay then.

Other people chose more tangible substances as their elemental matter. We met Heraclitus, who proposed fire as a candidate, in Chapter 1. Thales, who we met in Chapter 4, favored water because it took many forms, while the philosopher Anaximenes declared air to be the purest material, and so on.

It was a man named Empedocles who brought order to all the squabbling in the fifth century BCE. Rather than backing any of the other thinkers, he took the diplomatic approach and suggested that maybe everyone was right. Perhaps there wasn't just one element but several.[9] Empedocles's periodic table would have looked like this:

W Water It's wet	F Fire It's hot
E Earth It's brown	A Air It's airy

This surprisingly simple solution ended the arguments and everyone was happy. Thales could keep his water, Anaximenes his air, Heraclitus his fire, and Pythagoras was dead in a bean field so nobody cared what he thought.

Nowadays, some people still think of these substances as being elemental but there is really no justification for this. They were chosen for peace-keeping politics rather than accurate knowledge, although sadly a lie can remain popular if people like it and it is easy.

THE TABLE MAKES ITS DEBUT

Once Antoine Lavoisier discovered that air was a nitrogen/oxygen mixture and water was a hydrogen-oxygen compound, scientists abandoned Empedocles's four-element idea and began burning or dissolving everything they could lay their hands on to obtain the true elements.

By 1789, a lot of new ones had been discovered so Lavoisier gathered all the information and published a complete list, totaling thirty-three elements in all.[10]

He put them in four categories: gases, which were invisible but occupied space; metals, which were shiny and burned in oxygen; non-metals, which could be used to make acids; and earths, which didn't fit the category of metallic or acid-making.

Lavoisier's table was the first not to be based on guesswork or gut feeling and it looked like this:

Gases	Metals	Non-metals	Earths
Light*	Antimony	Phosphorus	Chalk*
Heat*	Silver	Sulfur	Magnesia*
Oxygen	Arsenic	Carbon	Baryte*
Nitrogen	Bismuth	Muriatic*	Alumine*
Hydrogen	Cobalt	Fluoric*	Silice*
	Copper	Boric*	
	Tin		
	Iron		
	Manganese		
	Mercury		
	Molybdenum		
	Nickel		
	Gold		
	Platinum		
	Lead		
	Tungsten		
	Zinc		

The substances indicated with an asterisk were later discovered not to be elements but for a first attempt his table was pretty good.

Other chemists had their own methods of grouping things, of course. The German chemist Johann Döbereiner grouped elements into families of three based on how similarly they behaved. The metals lithium, sodium, and potassium behave identically, for instance. They react violently with water, tarnish in air, and can be sliced with a knife (if you've never had the joy of cutting a piece of lithium metal, it feels like ice cream straight from the freezer).

A similar observation worked for sulfur, selenium, and tellurium. All three were powdery solids that reacted with oxygen to produce strong-smelling compounds. Döbereiner called these groups triads, but there was no apparent reason for the patterns.[11] The finished table of elements would have to somehow explain these mysteries.

A MUSICAL INTERLUDE

The most famous stab at a periodic table, before the one which actually worked, was a doomed attempt by the Englishman John Newlands in 1863.[12] Methods had already been devised to measure the weights of atoms pioneered by Swedish chemist Jöns Berzelius (who also introduced the element symbols we use today)[13] so Newlands obtained the data and wrote a list of the elements in order of ascending mass. As he did so, he discovered that the elements *almost* followed a cyclic pattern the way musical notes do.

In Western music theory, there are only seven principal notes. If you start at any particular tone and play up the scale you'll discover that the eighth note is identical to the first, just a higher version. Note nine is a higher version of note two and so on. One complete set of notes is called an octave and the notes spiral up and up until the human ear can no longer catch them.

John Newlands applied the same logic to his table of elements, claiming there were seven categories that repeated over and over as we got to higher masses. The first seven elements made the first row, while the eighth element would be the first entry on row two, having similar properties to element 1 directly above it.

He called the seven columns of his table "families" and the eight rows "periods," meaning something that repeats regularly. Thus, John Newlands introduced the idea of elements being "periodic."[14]

The idea of periods turned out to have some truth to it, but his table had one minor flaw, which can sometimes prove inconvenient for a hypothesis: it was wrong.

At the time Newlands composed his table (pun very much intended), there were sixty-three elements known, which didn't fit into an eight by seven grid. So rather than adding an extra column or abandoning the octaves idea, Newlands shoved a bunch of elements into the same grid squares.

The metallic element cobalt, for instance, having the audacity to exist, nudged later elements out of their correct families, which didn't match the hypothesis. Newlands decided that cobalt and nickel were therefore the same element.

They aren't. (Although, fun fact, both get their names from German sprites, Kobold and Nickel.)

Newlands knew these elements weren't the same as each other but this kept his table neat, so best not to worry about it. He then had to do the same thing with awkward vanadium and again with lanthanum. In doing so, Newlands fudged the data to fit his idea. We have a word for that in science: cheating.

It would be like claiming there were three types of animal: cows, goldfish, and pigeons—then when someone shows you a tiger you decide it's a cow really and put it in the same column.

Newlands also cherry-picked the elemental features. Cobalt is a lustrous metal with magnetic properties but his table aligned it with fluorine, chlorine, and hydrogen, all reactive gases. Newlands was happy to point out that chlorine, hydrogen, and fluorine belonged together but ignored the fact that cobalt didn't.

As a scientist, your job is to recognize when your hypothesis has failed. If nature says your idea is wrong then you get a new idea, you don't tell nature what to do.

As a result, Newlands's table was rejected by the scientific community of the day, although the story does have a happy ending. Every scientist has published a dodgy idea at some point, so scientists are a forgiving bunch who try not to hold grudges. If one idea turns out to be wrong, your others are still given a fair hearing. It's useful to have that approach because, although Newlands's octave hypothesis was wrong, his idea of periodic repetition turned out to be on the money. Elements do obey a cyclic pattern but a much more complicated one than he had assumed. He was, for this realization, awarded the Davy Medal for Chemistry by the Royal Society in 1887.

THE DREAMER

Dmitri Mendeleev was born in Siberia in 1834, the youngest of probably thirteen children (historians can't agree on the number, but I'm sure his parents knew).

When his father went blind, Dmitri supported the family financially by tutoring science and, according to those who saw him in action, he was a fantastic communicator, full of passion and enthusiasm for both the subject and the art of explanation.

At the age of fifteen, his mother decided he needed a higher education and took him across Russia on foot, applying to as many universities as they could along the way. The expedition took close to a year and sadly her health worsened as the months drew on. She died when they reached St. Petersburg, but lived long enough to see her son get admitted to study joint-honors in chemistry and teaching at St. Petersburg State University.

She would have been proud of his accomplishments, as he soon became one of the most outstanding chemists in Russia, with a reputation for writing huge textbooks from memory in a matter of months, and helping establish the country's first oil refinery in Tutayevsky.

He was also an imposing character, who shaved his beard once a year and had fiery clashes with other students and professors. His greatest contribution to science though was creating the very first periodic table that actually worked.[15]

A few days prior to his breakthrough, Mendeleev made a deck of playing cards with elements instead of suits on their faces. He invented a version of solitaire based on chemical properties and hoped it would help him discover a deep pattern about their organization.

According to his friend Alexander Inostrancev, Mendeleev had been awake for three days and nights playing the game when he finally collapsed from exhaustion on the afternoon of February 17, 1869.

Mendeleev fell asleep surrounded by his playing cards and had the most vivid dream of his life. In the dream, he saw the playing cards dancing before his eyes and dropping into place perfectly, revealing the pattern for which he had been searching.[16] The elements did follow a cycle, but nobody had figured it out because there were still elements missing!

Up until then, people had been discovering elements at random and grouping them based on color, reactivity, conductivity, thermal properties, and anything else you could name.

Mendeleev realized that the elements were arranged in a sequence of increasing mass but that some were still hidden inside rocks. The elements that seemed to be in the wrong place weren't: they were just next to elements that were unknown to chemists of the day.

Element 32 hadn't been isolated yet, nor had 61 or 72. If we assumed Mendeleev's law of increasing integers worked, we should find elements that matched those values and, sure enough, germanium, promethium, and hafnium were eventually identified and slotted into their respective gaps.

THIS WAY MADNESS LIES

By 1932, we knew that elements were made of atoms, themselves made from protons, neutrons, and electrons. But if every atom was made of the same three particles, why were they so different from each other?

Take element 35, bromine. It's a thick, mauve liquid that sets fire to metal and corrodes human skin. The next element is number 36, krypton. That's a harmless, invisible gas with no odor or reactivity. The only difference is that krypton has one extra proton/electron than bromine, so why don't they behave similarly?

And what can we make of Döbereiner's triads? Elements 29, 47, and 79 are copper, silver, and gold—all malleable metals with a lustrous finish. Why do those three numbers in particular end up with the same properties?

Why is element 4 a shiny solid while element 5 is a brown powder? Why is element 9 one of the most reactive known to man but element 10 one of the least? Why do elements 11 to 14 conduct electricity while elements 15 to 18 do not?

Any attempt to find order resulted in failure and a hypothesis has to account for all evidence, not just a convenient portion of it. If we couldn't use the proton/neutron/electron model to account for the differences in behavior then we would have to abandon it.

The only conceivable explanation was that although each atom was made of the same three particles, they were somehow arranged differently in space. Democritus had already suggested that atoms came in different shapes (fire atoms were spherical, which allowed them to move easily, while "bitterness" atoms were sharp and jagged). Could he have been on the right track?

The answer finally came when physicists discovered one of the most important theories in modern science, the one that gave the periodic table its final form. Quantum mechanics.

CHAPTER SIX

Quantum Mechanics Saves the Day

QUANTUM CRASH COURSE

Quantum mechanics is infamous. Everyone has heard about it and its reputation for being weird (a reputation that is well deserved, by the way). However, in recent years, some of the vocabulary has been hijacked by spiritualists to mean all sorts of unrelated things, which sadly confuses the issue. Don't misunderstand me; there's nothing wrong with talking about spirituality but repurposing words from quantum mechanics to mean something else is unhelpful. So we'll tread carefully.

The first thing to say is that quantum mechanics is not one idea but a sophisticated collection of theories that explain the world at its smallest level. The behavior of electrons, the nucleus, light, and their interactions are all explained by quantum mechanics so it is of great importance to chemistry.

Covering it in detail would take a separate book entirely so we'll limit the discussion to the part, developed by Austrian physicist Erwin Schrödinger, that helped build the periodic table.

Schrödinger caused a lot of discomfort during his life and was politely asked to leave a number of universities and institutions. This wasn't because of his academic achievements, which were outstanding. It was because he lived in a three-way relationship with his wife Annemarie and their girlfriend Hilde. He also wore a lot of bow ties. Scandalous.

Schrödinger's most important contribution to science is called the Schrödinger wave equation. It's the equation that tamed the periodic table and explains why elements behave the way they do. It looks like this:

$$H|\Psi\rangle = i\hbar \frac{\partial |\Psi\rangle}{\partial t}$$

I know equations can sometimes put people off but this one is vital to the story, so we can't just brush it under a rug. I've included a short explanation of what it means in Appendix III if you're feeling adventurous but don't worry, we can still understand what the equation does without having to go into any mathematical detail.

Nobody is sure how Schrödinger came up with his equation because there are no clear records of him deriving it. Some claim he simply woke up one morning, went downstairs, and wrote it based on gut feeling. It was only later that it was tested and proven correct.

What the equation does is tell us where electrons are likely to be as they zip about the nucleus. You start by taking the electron's properties (things like its mass, velocity, etc.) and then figure out how much attraction there is from the protons of whichever atom you want to describe.

By solving the equation for a given atom we can map out a three-dimensional region of where electrons are going to be and what patterns they will trace out in space.

When we do this, we find that electrons don't move in circular orbits at all. They surround the nucleus in regions that come in a variety of shapes, the same way animals inhabit different-shaped enclosures at a zoo. We call these regions "orbitals" or sometimes when we're being lazy, "electron clouds."

Some electrons hang out in spherical orbitals while others occupy a dumbbell-shaped region protruding from the top and bottom of the atom. Each orbital can hold up to two electrons, so the more electrons you have on your atom the more orbitals end up being used and the more extravagant your atomic shape becomes.

The reason certain orbital shapes arise is because electron movements are sort of wavy. They don't move in simple lines like marbles but seem to ripple as they travel from one point to another. Since ripples can only come in certain shapes (you can't have half a wave, for example) so do the electron orbitals.

A boron atom, which has five electrons, will distribute them into orbitals shaped like the diagram on the left on page 65. Carbon, however, has six electrons so a new orbital shape gets introduced and the atom looks like the diagram on the right.

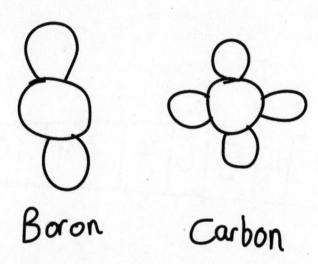

The fact that different atoms come in different shapes explains why they have difficult chemical behaviors. They stack together differently, bond at different angles, fit into different spaces, and so on.

Solving the Schrödinger equation for a particular element explains why it can be different to the element next door. Just because they have a similar number of electrons doesn't mean they will have the same shape. It also answers the question every student asks when they see the periodic table for the first time.

WHY IS IT *THAT* SHAPE?

A table is supposed to be a neat rectangle with columns and rows. You know, like the one Lavoisier created. The periodic table we use today looks like a chimp accidentally vacuumed up a computer keyboard and tried to glue it back together with silly putty. It doesn't look table-ish at all. So, who came up with the design and why did everyone else say "Yup, looks good to me"?

The man to thank is one Alfred Werner, the Swiss Nobel Prize–winning chemist who published a short article in 1905 with the catchy title "Contribution to the development of a periodic system."[1] It was here that the periodic table first took shape.

Let's consider the first ten elements. Actually, let's not. Let's ignore elements 1 and 2 and begin with element 3. (I'll explain in a moment.)

We could line the elements up in a nice long row and be done with it:

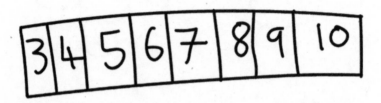

But now we can do better thanks to the Schrödinger equation. The first two elements of this row put their electrons into spherical orbitals while the next six go into dumbbell-shaped orbitals. This means we can split the line like so:

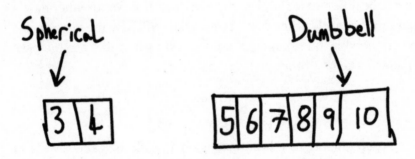

The next eight elements have the same orbital shapes. The atoms will be bigger but they will otherwise have very similar chemistry. To represent this, we use Newlands's periodic idea and add a second row to our table, still dividing into two blocks:

 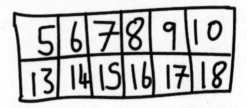

Each column of elements represents a particular orbital shape. The only difference is that, as we go down, the orbitals get larger.

When we get to element 21 a new shape gets introduced (quantum mechanics is like that). The outer electrons of the atom of this element, scandium, up to element 30, zinc, are shaped like bundles of balloons rather than dumbbells so we need to introduce a new block to the table. Element 31 goes back to the dumbbell shape and so our table now looks like this:

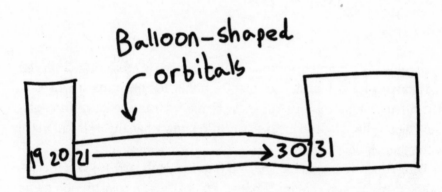

It's a bit irritating that nature insists on introducing weird orbitals when we get to larger elements, but that's why the table is an awkward shape. It's because nature is.

Now, if you read from left to right across a row (period) you're reading in ascending proton number, while the column (group) tells you what shape the atoms are going to have. Reach the end of one period and you just go on down to the next one.

When Alfred Werner included all the known elements and orbital shapes, the table ended up like this:

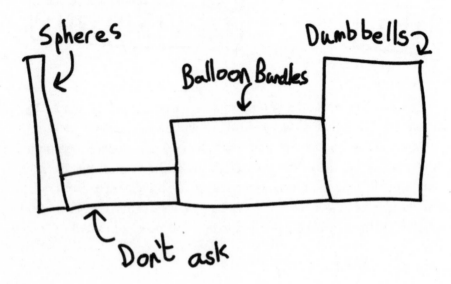

Suppose you wanted to know about iodine. By counting from left to right you learn that it is element number 53, meaning it will have fifty-three protons and fifty-three electrons. You can see it's in the right-hand block (dumbbell shaped) so you also know what angles it will make with other atoms.

Directly above it are chlorine and fluorine, both colorful non-metals. Iodine is in the same column so it will probably be a colorful non-metal too, but with a higher density as it's on a lower period. Sure enough, we discover that these are exactly the properties of iodine.

You can even use the periodic table to predict the properties of elements nobody has ever seen. Directly below iodine is astatine, the rarest element in the Earth's crust (less than 1 g exists on the whole planet), but if we had a sample it would probably behave like a denser version of iodine. God bless quantum mechanics.

ARCHITECTURAL SIMPLICITY

You probably know from your periodic table T-shirt, mouse pad, shower curtain, pencil case, and notebook (I own all of these and assume everyone else does too) that the above pictures aren't quite there yet.

This fully expanded version of the table is rather cumbersome, so for simplicity we take one of the blocks, scooch it down to the bottom, and slide the others in to meet each other.

This form of the periodic table was proposed by Glenn Seaborg in

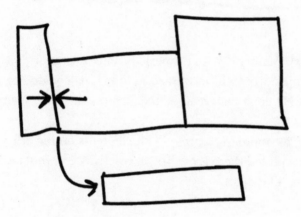

1945 and soon became the standard thanks to its simplicity and the fact that Seaborg did a lot of work to popularize science.[2] But obviously we've missed out elements 1 and 2.

Hydrogen and helium are both spherical atoms meaning they belong in groups 1 and 2 respectively:

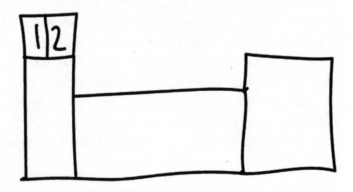

That's what Harvey White did with his periodic table design in 1934 and it's what Schrödinger would have wanted.[3] Unfortunately, due to their small size, H and He don't behave quite the same as the other elements in that block.

They actually have more in common with elements on the other side of the table so if we placed them according to reactivity we would end up with things looking like this:

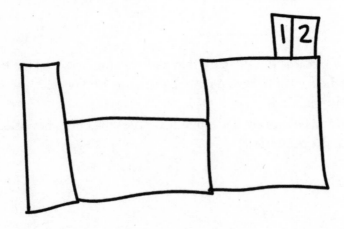

This is what Ernst Riesenfeld did with his periodic table in 1928 (and what Mendeleev would have wanted).[4]

Glenn Seaborg couldn't make his mind up about where to put these two fiddly elements, sometimes drawing them on one side and sometimes the other (and briefly putting hydrogen in two groups in 1945).[5]

Eventually, the general agreement was to acknowledge both the electron-orbital work of Schrödinger and the chemical-properties work of Mendeleev.

So we split them up and put them at each end of the table, one for each scientist. It's not logical but it's a nice tribute to the two men who built it. And *voilà*, we have ourselves a periodic table.

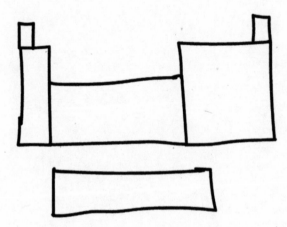

CHAPTER SEVEN

Things that Go Boom

THE MOST EXPLOSIVE EXPLOSIVE

With the exception of nuclear weapons, most explosives work in the same way. First, a material is synthesized, which is highly unstable. In chemical terms this means it will fall apart given the chance. Second, the material is provoked, giving it the chance to break up and rearrange into stable substances. During this rearrangement a whack-ton of energy gets released (whack-ton being the technical term) in the form of light and heat.

In addition, a small amount of solid or liquid explosive will expand rapidly to become a large volume of gas. This sudden expansion combined with lots of heat and light is what we call an explosion.

Some substances are so unstable that even a little bit of agitation will cause the reaction to begin. Gunpowder only needs a candle flame to decompose, while TNT requires nothing more than a spark. You can even buy bang snaps, children's toys consisting of small paper parcels filled with silver fulminate, a chemical that explodes when struck. The parcel is thrown at the ground and the impact causes a loud snap.

Fireworks work on a similar principle. A powdered metal is launched into the air and once the fuse within has burned, a detonator gel will trigger, converting it into a gas. Each fleck of powder is sprayed outward as the gas expands, becoming so hot they start reacting with the atmospheric oxygen, giving us sparks.

We've already seen in Chapter 4 that different elements give out different types of light, so by picking particular metals we get particular spark colors. Sodium will turn yellow, barium goes green, copper blue, and strontium brick red. Purple fireworks are famously difficult to achieve and usually involve a mixture of copper and strontium together.

All explosives rely on the chemicals within being unstable, and the most unstable chemical ever created is called azidoazide azide, synthesized in 2011 by Thomas Klapötke. Although why anyone would want to make this chemical is beyond me.

It contains fourteen nitrogen atoms and two carbon atoms clumped into branches around a tight ring, all squished together without much room. The bonds between the atoms are so strained that they spring apart under any circumstance.

When Klapötke tried to dissolve the chemical in water, it exploded. When he tried to move it across his lab, it exploded. When he breathed in its general direction, it exploded. It even exploded when infrared light (the kind emitted from a TV remote control) was shone at it.[1]

The best explosives, of course, are those that detonate on cue. They have to be stable enough to be moved, but unstable enough to still detonate. Azidoazide azide would be a poor choice, for the same reasons chlorine trifluoride was a terrible choice of rocket fuel. You're better off with good old dynamite, invented by none other than Alfred Nobel.

THE MERCHANT OF DEATH IS DEAD

When somebody dies, we tend to say nice things about them because it's taboo to speak ill of the dead. Such was not the case on February 12, 1888, when Alfred Nobel's obituary was published. A French newspaper allegedly ran the headline "The Merchant of Death is Dead" and went on to say, "Dr. Alfred Nobel, who became rich by finding ways to kill more people faster than ever before, died yesterday."[2]

A lot of people were not happy with a newspaper remembering the great scientist in such a way. That included Alfred Nobel himself, who managed to read his obituary owing to the fact that he was not really dead. The story goes that Alfred's brother Ludvig had died and the newspaper mistook the two brothers, prompting them to publish their powerful attack.

Nobel was a very talented chemist who had invented dynamite twenty-one years earlier. Initially he intended it for use in mining, but it had obvious military applications. Apparently, realizing what his legacy was, Nobel decided to alter his will and left his considerable fortune (then over thirty-one million Swedish kronor, equivalent to nearly eighty million US dollars today) as prize money to people who did things "for the greatest benefit of mankind." The prizes were to be awarded for achievements in the three sciences, literature, and the promotion of peace: the Nobel Prizes.[3]

The newspaper that ran the obituary is reported to be the *Idiotie Quotidienne* and I have desperately tried to find documentation to confirm its existence, but sadly I can find none.[4] The fact the newspaper's title translates to the Daily Idiocy might be a clue that the story is a hoax and indeed some of Nobel's biographers have dismissed it as a persistent rumor.[5]

It may be an apocryphal morality tale or perhaps an embellishment of Nobel's reaction to his brother's death. Whether it's true or not, the Nobel Prizes are still considered the most prestigious awards it is possible to receive in science. The prize money is substantial, numbering in the millions, and it comes from Nobel's estate built entirely on dynamite. Not literally, obviously. That would be stupid.

The way dynamite works is simple. You take a large amount of silicon-based rock powder and soak it in a chemical called nitroglycerine. Pack this into a tube and stick a fuse in the end. As the fuse burns, the heat is transferred to the nitroglycerine-soaked powder and—boom!

Nitroglycerine is one of the unstable compounds I mentioned earlier. Composed of carbon, nitrogen, oxygen, and hydrogen atoms, it is a chemical that will self-react, i.e. one nitroglycerine particle will react with another to produce a bunch of gases, mostly carbon dioxide and water.

These gases expand to over twelve hundred times their original volume and reach a temperature of 5,000°C. The reaction is also fast with the expansion and heating taking place in under a microsecond.

All of this comes down to the question we're going to answer in this chapter—why do chemical reactions happen at all? What do we mean when we say a chemical is unstable and how do atoms bond in the first place? To make sense of it all, we'll need to dive a little deeper into the quantum ocean.

GIVING CHEMISTRY A BAD NAME

The word chemistry comes from alchemy, but a better name for the subject would be electronics, because chemical reactions are all about electrons. The nucleus of an atom is tiny compared to the overall radius so it's the electrons on the outside that are interacting with everything.

And electrons are always on the move. Were an electron to cease moving it would simultaneously cease to exist because movement, like the charge it possesses, is part of an electron's identity. A stationary electron exists no more than a four-sided triangle.

So, if we take movement as a given, there are only two things an atomic electron can really do. It can move outward away from the nucleus or it can move inward toward it. These two behaviors underpin almost every chemical reaction you'll meet.

Let's revisit the concept of orbitals that we got from the Schrödinger equation. They're the regions around a nucleus where electrons spend their time.

Orbitals are the permitted electron territories, but electrons aren't confined to live their entire lives in the same one. They can hop around. When an electron jumps from one orbital to another it's called a "quantum leap" and it can happen between any two orbitals, even ones that are empty.

Obviously, electrons prefer to occupy an orbital near the nucleus because it carries the opposite charge but they don't always get their way. If the innermost orbitals are inhabited, other electrons have to make do with being further out.

An atom is really a hustling, bustling place where the orbitals nearest the nucleus are considered prime real estate and every electron wants to move in. If one of the inner electrons happens to vacate their orbital for some reason, an outer electron will quantum leap to replace it.

These quantum leaps don't happen at random, though. The rule that ultimately accounts for chemistry is as follows: if an electron absorbs a beam of light it gets bumped to an outer orbital and if it emits a beam of light it drops to an inner one.

Some types of light have more ability to promote electrons while others have less. A blue beam can promote an electron to a far-out orbital while red light might only nudge it up by one level. In the same way, electrons from far-out orbitals have the ability to release blue light when they drop, while electrons already near the nucleus might only release red.

This is how the fireworks and spectroscopy we've already mentioned work. Every atom has a unique orbital arrangement so every atom emits or absorbs a unique light spectrum. When the electrons start jumping from orbital to orbital, the distance they jump determines what kind of light gets emitted or absorbed depending on which way they're traveling.

The question everyone usually asks at this point is why electrons absorb or emit light in the first place. I'm afraid the answer is because that's just the way nature is. It's just one of the fundamental laws that were established during the big-bang expansion. The same way balls roll downhill as they obey the laws of gravity, electrons release and absorb light as they obey the laws of quantum mechanics.

ABILITY AND STABILITY

We've talked about some beams of light having more ability to promote electrons than others, and in science we sometimes replace the word ability with the word energy. I've mostly avoided using it up until now because it's a word fraught with difficulty and misconception.

People talk about energy as if it were a thing being transferred from place to place, but it isn't really. You can't hold a lump of energy but a lump of matter can possess the ability to bash into things or the ability to explode, i.e. it can possess energy.

In the context of quantum chemistry, energy means "how capable a beam of light is to push an electron into a higher orbital." You'll sometimes hear scientists say that electrons in outer orbitals have "absorbed energy" and that this energy gets released when they drop. This is convenient shorthand but we have to be clear: it is light that gets absorbed and emitted. Light has the ability to promote electrons and therefore possesses energy, but energy is not an actual thing.

The opposite of ability is what we mean by "stability" and it's a measure of how much energy an electron has lost when it drops down, or how reluctant it is to shift upward from its present orbital.

An electron from an inner orbital, close to the nucleus, is less willing to change because it is happy where it is. We describe it as being chemically "stable." An electron in a higher orbital with a lot of energy (ability to release light) is very unstable, however, because it is not happy and will change given the chance.

The diagram below shows what happens when an electron absorbs a beam of light. It jumps from a low-energy orbital to a high-energy orbital, becoming unstable.

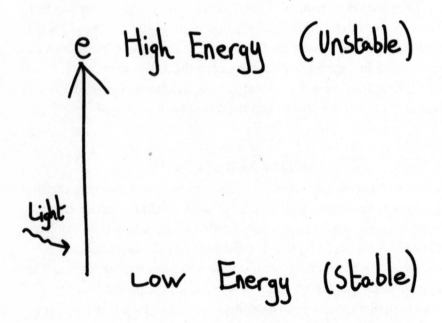

The next diagram shows the reverse process. This is a high-energy electron dropping down to a more stable orbital. The only difference is that light is being emitted here rather than absorbed.

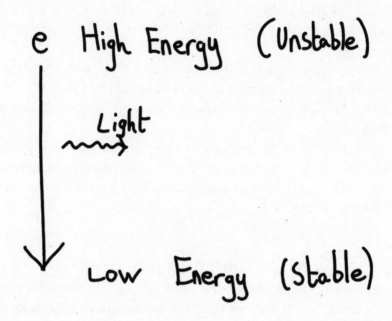

Ability and stability are always at odds with each other and govern an electron's reactive behavior. Gaining energy means losing stability and vice versa. This trade-off between ability and stability is what determines whether a reaction will happen or not.

SHAKE, RATTLE, AND ROLL

Different kinds of light will produce different kinds of effect on an atom. Infrared light, which is too low in energy to interact with the electrons in our eyes so we can't see it, will cause the orbitals themselves to stretch and twist rather than shunting electrons between them. Microwaves do something similar except they cause the atom to spin, rather than twist and bend.

If you beam atoms with infrared or microwave light the result is that the atoms start dancing around and bashing into each other, exchanging energy. Ultimately this happens via the same light-transfer mechanism (electrons on one atom release light to electrons on the other, promoting them to a higher orbital/making the atom twist or spin more) but it's quicker and more convenient to talk about atoms colliding and transferring energy.

These twists and spins of the atoms are what we call heat and it's why you feel warm when infrared or microwave light hits your skin. It also provides the basis of microwave ovens by causing the water inside a piece of food to jiggle.

Obviously, the hotter a sample of chemical, the more likely it is the atoms will bump into each other and trigger orbital rearrangements/twists/spins. Or, put another way, heating most reactions tends to make them happen faster.

UNITED WE FALL

Imagine being an electron tethered to an atom's nucleus. If another atom approaches, its nucleus can draw you toward it at the same time. If the pull is strong enough you can be dragged into a position halfway between both nuclei and you are no longer occupying an atomic orbital but a "molecular orbital." A molecular orbital is known by a more common name: a chemical bond.

If the molecular orbitals are at lower energy than the atomic orbitals with which we started, then electrons on two approaching atoms can drop into a molecular orbital together, releasing light as they go. A bond between the atoms is formed and we have carried out a chemical reaction.

High energy
atoms

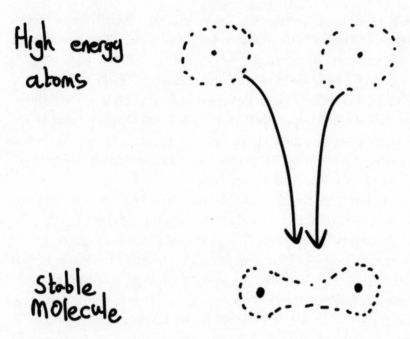

Stable
molecule

In the case of the hydrogen/oxygen reaction, the hydrogen and oxygen units are floating freely, but when they react the electrons on each atom slot into molecular orbitals, linking things together and forming a bonded H_2O molecule.

All the energy from these dropping electrons is released as light both visible and infrared (heat), creating the explosions first seen by Henry Cavendish.

It doesn't have to be atoms with which we start either, it can be molecules. The bonds of a nitroglycerine molecule are at very high energy so they gladly break down into molecules with more stable orbitals, like carbon dioxide and water. A lot of energy gets kicked out in the process as all the electrons drop and we see the result as an explosion.

LET'S GET IT STARTED

The basis of chemistry is simple: start with one set of orbitals and finish with another. Prizing your original molecules apart, however, can be difficult. Electrons in a molecular orbital don't necessarily know there's a better deal to be had, so we have to give them a kick of energy in order to let them fall into the arrangement we want.

An analogy would be to picture a coat hanging on a coat hook. The coat will sit there until the end of time, even though it would achieve greater stability by dropping to the floor. That won't happen because you have to put energy into the system first. It's only when you lift the coat up a few centimeters, freeing it from the hook, that you give it the option of falling into a more stable configuration.

Electrons are exactly the same. We need to excite them first and get them out of their orbitals before they can drop into new ones.

A stable molecule like water can be thought of as having a coat hook several meters long. You'd have to get on a ladder and lift the coat all that distance to get it free. And once you let go, it would probably just fall right back onto the hook again. That's why water reacts with hardly anything.

Nitroglycerine, on the other hand, is like a coat hook a few millimeters long, positioned over a cliff. A tiny nudge (say, from a burning fuse) is enough to get the electrons out of their orbitals, and the subsequent energy drop is enormous.

Or you could think of it like a LEGO® model. If you want to make something new, you have to put energy in and separate the blocks. It's only when everything is broken down into its constituent parts that you can form something else.

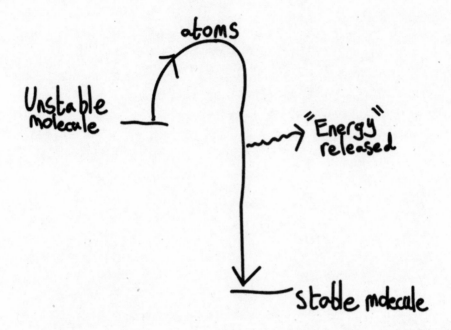

Whatever the reaction is, chemistry is about persuading the electrons to jump out of their starting orbitals and into the ones you want. How hot do you need to get it? What shape does your starting molecule need to be? What by-products do you get? What do you do if your reaction doesn't work? How many molecules will rearrange and how many will fall back into their original positions?

Although a myriad of complexities can arise in the lab, the overall premise is simple. Push the electrons up and let them drop down.

CHAPTER EIGHT

The Alchemist's Dream

THE MOST EXPENSIVE ELEMENT YOU'VE NEVER HEARD OF

On April 3, 2017, the Pink Star diamond was sold at auction to Chow Tai Fook Enterprises for a cool seventy-one million dollars.[1] At the time of writing this is the largest sum of money ever paid for a gemstone.

For perspective, the Hope diamond was sold in 1908 to Selim Habib on behalf of the Sultan of Turkey for $200,000, then resold in 1911 to Evalyn McLean for $154,000. In 1958, it was gifted to the Smithsonian Institution in Washington, DC, insured for one million dollars and rumored to be worth even more today.[2]

Diamonds are pure carbon so perhaps it would be fair to call carbon one of the most expensive elements on the table. Then again, charcoal, which is also made from pure carbon, retails for a few dollars at any supermarket. So perhaps it's one of the cheapest.

We treat gold as a more valuable metal than silver but in the 1890s the winner of an Olympic event was presented with a silver medal rather than a gold one. Record companies reward artists with a platinum album as their highest accolade, but platinum sells for fifteen dollars less per troy ounce than gold on the open market.

Rhodium and palladium, used to make catalytic converters in cars, currently have a similar value to platinum but enjoyed a brief spike in 2008 when their value increased ten-fold, making them more valuable than gold for a month. Things are only worth as much as someone is willing to pay for them, and the elements are no different.

Plutonium is one of the most expensive materials on Earth for obvious reasons, with a value of $11,000 per gram (according to the US Department of Energy), and it's often reported as being the most expensive element.[3] But there is one other, rarely discussed, that outranks it. Californium, element number 98, is used as a starting agent in nuclear reactors and sells for a titanic twenty-seven million dollars per gram.[4] The Pink Star diamond weighs about 12 g, meaning californium is over five times more expensive gram for gram.

What makes it so pricey is that californium does not occur in nature. It's an element we have to make for ourselves.

ALL OF THEM WITCHES

Before the discovery of phosphorus and the fire experiments of the eighteenth century, chemical research was a mess. Armed with a mixture of Judaeo-Christian symbolism, ancient fairy tales, and the works of a Persian writer named Jabir ibn Hayyan, rigorous testing of chemicals was ignored and fact was mixed with superstition.

The resulting field was called *alchemy*, an Arabic-derived term which comes from the Greek *chemia*, meaning black magic. Nobody was trying to find substances that were elemental during that period. Instead, they were trying to find substances they more or less made up.

"Alkahest" was one, thought to be the ultimate acid capable of dissolving anything. The "elixir of life" was another, thought to prevent the onset of death, and the "panacea" was yet another, thought to be a medicine capable of curing all illnesses.[5]

Above all, though, the goal of the alchemists was to generate a material called "philosopher's stone," which could turn other metals into gold. Nobody knows who came up with the idea of philosopher's stone but rumors of its existence had been circulating since the thirteenth century.

The author of a medieval encyclopedia, Vincent of Beauvais, claimed that God had imparted knowledge of "transmutation" to Adam, who passed it on to Noah and so forth. His source for this seems to have been his own imagination although *The Book of Sydrac*, an anonymous thirteenth-century text, tells a similar story, so it was obviously a common idea at the time.[6]

One of the earliest recorded references to the phrase "philosopher's stone" is in a 1610 play called *The Alchemist* by Ben Jonson, which suggests that Adam was told how to make the fabled substance.[7] After he got kicked out of Eden, presumably he forgot the recipe. Nice going, Adam! First you lose a rib, then you lose the philosopher's stone recipe. What's next? Your second-born son?

Alchemy did give us knowledge about various chemical reactions, not to mention Brandt's discovery of phosphorus, but it had no structure to it and there was more guesswork than anything else.

The problem with trying to turn one element into another is that an element's identity is determined by the number of protons in its nucleus and changing that isn't a simple matter of mixing things in a test tube.

As we saw in the previous chapter, chemistry is all about manipulating electrons. The nucleus is too small and hidden for us to have any impact on it. Simply put, electrons can dance to any tune you want but if the nucleus remains untouched the element remains the same.

And yet suns are constantly transmuting hydrogen into helium so there is obviously no law of science that forbids it from happening. To mimic the technique here on Earth would take superhuman powers. Speaking of which . . .

THE ORIGIN OF SUPERHEROES

Peter Parker got his Spider-Man powers when he was bitten by a radioactive spider and his DNA became irreparably altered. Bruce Banner got caught in the blast of an atom bomb and was belted by radioactive gamma rays, turning him into the Hulk. The Fantastic Four were caught in a storm of radioactive cosmic rays, Daredevil was splattered with radioactive waste, and Jean Grey of the *X-Men* (in the original storyline) released her telekinetic potential from flying a shuttle through a radioactive solar storm.[8]

Radioactivity has obviously given us much to be thankful for, but it also created Godzilla and an uncountable number of giant insects during the 1950s so we should probably treat it with caution.[9] Nevertheless, it was through radioactivity that humankind was finally able to transmute one element into another, so we need to get acquainted with it.

The phenomenon was discovered by accident in 1896 by the French physicist Henri Becquerel. Becquerel had been planning to do some experiments with photographic plates but on the day of his tests the sky was overcast, so he put them in his drawer.

Two days later when he got them back out, the plates had somehow been impregnated with the image of a copper cross lying next to them. Apparently, something in the drawer had taken a photograph. The only other object present was a jar of potassium uranyl sulfate solution on the other side of the cross so Becquerel decided it had to be the culprit.

While a photographic plate is best activated by sunlight, any high-energy beam will cause it to undergo a change. Potassium and sulfate particles don't emit beams, so logically it was coming from the other element in the liquid, uranium.

Invisible to the human eye, the uranium was apparently emitting something that altered the surface of the photographic plates. The cross had got in between them and, *voilà*, the first radiogram taken by a jar.

Soon after Becquerel's discovery, Marie Curie, the only person to win Nobel Prizes in two sciences, named the phenomenon *radioactivity* from the Greek *radius* (wheel spoke) and the Latin *aktinos* (ray).

With her husband Pierre, Marie discovered two more radioactive elements, which she named radium (for obvious reasons) and polonium after her home country of Poland. Sadly, both the Curies succumbed to illnesses caused by their exposure to radioactivity, which taught us something else—it's damaging to cells.

INHERENT INSTABILITY

As we learned in Chapter 3, the nucleus of an atom is an unstable design. While the protons hold electrons in place, they also repel each other, which requires neutrons to glue them together.

The Austrian-born scientist Lise Meitner figured out that once you got up to elements around the high eighties, this equilibrium would become unstable and the nucleus could fall apart. For this important discovery, Meitner did not win the Nobel Prize for physics. Her male lab partner did. But Element 109 was eventually named meitnerium after her, so she hasn't been snubbed completely.

As we ascend through the elements, proton numbers increase so the neutron numbers have to follow suit to keep things together. But there's a complication (isn't there always?). The repulsive force between protons has an infinite range but the glue force from the neutrons doesn't.

This means that in large atoms it's only a matter of time before repulsion wins, making them precarious structures. Larger atoms are fragile and left for long enough will break apart.

Blue-glowing actinium has a colossal nucleus of eighty-nine protons so, if you have a lump of it, around half will decay into something else within twenty years. Rubidium by contrast is much smaller, with only thirty-seven protons, and takes forty-nine billion years to decay by the same amount.

The nuclei these elements turn into tend to have peculiar numbers of neutrons, which the element doesn't normally have. These "daughter" particles can only be produced from radioactive decay, so if you measure the amount of mother and daughter nuclei in a rock, the ratio between them allows you to work out how much you had to start with and subsequently how long it's been around for.

It was using this technique that the American chemist Clair Patterson calculated the age of the Earth to be approximately 4.5 billion years old.[10]

BREAK IT DOWN

There are different ways a nucleus can decay. Sometimes the whole thing will split in what we call fission, but for reasons that aren't understood the most common thing to get ejected from a nucleus when it fractures is a bundle of two protons and two neutrons moving at tremendous speed.

These packets come hurtling out of their atoms at 15 million meters per second, and turn out to be the very same alpha particles Rutherford used in the gold-foil experiments.

When an alpha particle is emitted the nucleus left behind has lost two protons, changing its identity. Rutherford decided to use this to his advantage. Given the velocity of alpha particles, he proposed that if you shot them at another atom they could shatter its nucleus, turning it into something lighter.

By firing alpha particles through a highly pressurized container of nitrogen gas to increase the chances of collision, Rutherford was eventually able to bash nitrogen atoms apart, turning them into carbon, in 1919. His experiment made headlines because he had "split the atom" and achieved transmutation between elements. The long sought-after dream of the alchemist was not a mythical stone from the Garden of Eden: it was a gas chamber and an alpha-emitter.[11]

Turning lead into gold might not be possible through sacred incantation but, if you take an element like thallium, boil it to a gas, pressurize it, and fire alpha particles through the sample, one in every few thousand thallium atoms would be turned into gold.

BUILD IT UP

Alpha decay makes a certain amount of sense to the human mind because we can imagine something falling apart when repulsion overcomes attraction. There's another thing that can happen inside the nucleus, though, which can't be visualized so easily. Neutrons can turn into protons and spit out an electron as they do so.

There's a detailed account of how this happens in Appendix IV but it would take us way off track at this point. The best thing to do is cheat and think of a neutron as being a proton with an electron wrapped around it like a candy wrapper. If the electron is peeled off and discarded, the resulting particle will be a proton.

We call these streams of ejected electrons beta radiation and, unlike alpha decay, which only happens to heavy nuclei, the neutron/proton transformation can occur in any element. Some are more susceptible than others (those with more neutrons) but any atom is potentially beta radioactive.

$$\text{\textcircled{n}} \longrightarrow \text{\textcircled{p}} + e$$

Beta radiation

If we could persuade an element to turn one of its neutrons into a proton, we could obtain an element one number higher than that with which we started: Rutherford's process in reverse. But first, bananas.

BANANAS

Radioactive particles are charged and move at high speed, which means they destroy things in their path including the chemicals of your body.

If you become exposed to enough radioactive beams the DNA in your cells will fall apart and your body will disintegrate from the inside out. Usually, the fastest-growing parts (hair and nails) get affected first, which is why radiation sickness causes them to fall out. Then all sorts of lovely things happen like your skin peeling off, your teeth dropping out, and your innards gradually dissolving into a disordered mush.

In order to monitor the radioactivity to which a person is exposed, we measure the dosage in units called sieverts. A sievert is how much energy a radioactive beam is carrying, compared to the mass of the person it's entering.

There aren't clear figures on how many sieverts are dangerous to a human but roughly five-hundredths of a sievert per year is when things become problematic.[12] The most radioactive thing you're ever likely to come across is a dental X-ray or a mammogram scan, which delivers approximately 0.0004-hundredths of a sievert. A completely safe dose, in other words.

There is another unit that can be used to measure radioactive exposure: the banana.

The first thing to say here is that certain nuclei are more stable than others. By using the Schrödinger equation for protons and neutrons, we can obtain a list of especially stable nucleus values, which are genuinely called "magic numbers." There isn't an agreement on why this works—we just know that certain numbers of protons and neutrons are good and others are bad.

Potassium is a prime example. Most of the potassium atoms in the Universe are stable with nineteen protons and twenty neutrons, but around 0.012 percent have twenty-one neutrons instead, making them potassium-40, and this configuration happens to be unstable.

Potassium-40 will undergo beta decay readily so any sample of potassium will be emitting a very faint trickle of radioactivity, and the fruit that contains the most potassium is the humble banana.

Originally created as a joke in 1995 by Gary Mansfield at the Lawrence Livermore National Laboratory, the Banana Equivalent Dose (BED for short) calculates the amount of radioactivity one is likely to experience from eating a single banana and can be used to calculate the radioactivity of your food.[13]

Don't be alarmed, though: one BED comes to just under one-millionth of a sievert, so before you boycott bananas let's run the math.

If we assume five-hundredths of a sievert per year is lethal, you'd have to consume five thousand bananas fast enough for it to be dangerous. That's fourteen bananas a day. For a year.

If you really want to attempt this experiment, then I suggest you consult a doctor first. And probably a psychiatrist.

BACK TO PLAYING GOD

In 1940, the American chemist Dale Corson isolated element 85, astatine.[14] Predicted by Mendeleev's table, it was the last natural element to be discovered.

Slotting it into place gave us a periodic table that went all the way from 1 to 92 without any gaps. From hydrogen formed in the big bang to uranium formed in supernovae, every element was finally identified. But could we go further and generate our own with higher numbers?

In *Iron Man 2* Tony Stark is looking for an element to power his suit before it kills him from palladium poisoning. No appropriate metal exists, however, so in order to save the movie and defeat Mickey Rourke, he manufactures a new element using a UV-laser and his raw charismatic charm.[15]

We already know there are no elements missing from the periodic table so Stark's unnamed one is going to be made from huge atoms, and therefore massively radioactive. My screenplay for *Iron Man 3* centered around Tony Stark vomiting into a hospital bucket for two hours as radioactivity slowly destroyed his internal organs. For some reason, the script they eventually chose went a different way. Their loss.

In 1940, Edwin McMillan decided to pre-empt Tony Stark and make a new element for himself. He took a lump of uranium and fired a stream of high-energy neutrons at it until some were absorbed. A uranium nucleus can accept a neutron but doing so makes it unstable.

In order to lose some energy, one of the neutrons has to undergo a beta decay, kicking out an electron and converting itself into a proton. The uranium atom now has an extra proton in place of a neutron so it's not really uranium anymore. It's element 93.

For fourteen billion years the ninety-third element didn't exist in the Universe, and then suddenly, on Earth in 1940, it did.[16]

Uranium had been named after the planet Uranus, so McMillan named his element neptunium after the next planet in line. Later the same year, as part of the Manhattan Project, Glenn Seaborg managed to synthesize element 94. This one is a lot more stable than neptunium so, while neptunium was the first artificial element, Seaborg's could actually be made in chunks big enough to hold. It's a shiny metal with toxicity comparable to nerve gas, and Seaborg named it plutonium to keep with the planetary theme.[17]

Throughout the remainder of the Second World War and after, Seaborg went on to synthesize americium (element 95, named after America), curium (element 96, named after Marie Curie) and berkelium (element 97, named after Berkeley, California, where the research was carried out).

These experiments were highly classified as part of the war effort, but once it was over Seaborg was given permission to present his findings to the American Chemical Society on November 16, 1945. However, he accidentally spilled the beans five days earlier.

An avid popularizer of science, Seaborg was asked to appear on the children's radio show *Quiz Kids* and answer questions about physics. When an eleven-year-old boy named Richard Williams asked him if people would ever make new elements (not realizing he was talking to the world's leading expert on that very topic) Seaborg was unable to contain his excitement and blurted out the classified discoveries live on air, much to the annoyance of his superiors.[18]

Really though, can we blame him? He was surrounded by eager minds asking him everything he knew about his favorite topic. One might say he was in his element. I've been waiting for eight chapters to drop that joke.

COMPLETING THE TABLE

The periodic table is split into seven periods representing the seven orbital shells, and eighteen groups representing how many electrons occupy each one. As a result, the table has 118 spaces. With 92 occurring naturally, that gives us 26 blanks to fill.

Seaborg got lucky with his elements because they were all reasonably stable. Had he kept going he would have found things got a lot more difficult. Forcing nuclei to put on weight isn't easy because the larger they get, the more repulsion there is between protons.

The best approach is to take samples of an already large element and bombard it with smaller nuclei in the hope that they get absorbed. In 1950, californium was made by firing alpha particles at curium, and einsteinium and fermium were made in 1952 via a similar route.

We've also used this technique to create lower-numbered elements, which are normally rare in nature. Francium is the second scarcest element on the table (behind astatine) with approximately 30 g available in the Earth's crust. But if we fire an oxygen atom at a piece of gold we can generate it.

We can also create supplies of technetium, element 43, which has an unstable nucleus and doesn't normally last. It's worth doing because it makes up 80 percent of the world's medical tracers, injected into the body to track blood flow.

Making artificial elements is a precision operation, of course. Fire the nuclei too slow and they bounce off; go too fast and everything shatters. But over the last half century we have edged closer and closer to a full table.

We've fired carbon atoms at americium and curium to make mendelevium and nobelium. We've fired neon at einsteinium and made lawrencium. We've fired neon again at plutonium and made rutherfordium.

By the early 2000s, we had dubnium, seaborgium, bohrium, hassium, meitnerium, darmstadtium, roentgenium, copernicium, flerovium, and livermorium, leaving only four to be discovered. The missing numbers were 113, 115, 117, and 118.

At this point, the bottom-right of the periodic table looked like a row of punched teeth. Then, in November 2016, the International Union of Pure and Applied Chemistry announced the successful synthesis of nihonium, moscovium, tennessine, and finally oganesson. The periodic table was finally complete.[19]

Some of this might seem like pointless playing around but many of these artificial elements can be useful. You probably have a sample of americium, element 95, in your home right now. At least, I hope you do.

Americium emits alpha particles constantly so if you put it in an open circuit, the charged particles can fly across a gap to a receiver and complete the circuit without wires. When flecks of smoke or dust float into this gap, the alpha stream gets blocked and an alarm triggers. This is how your smoke-detector works.

THE END OF EVERYTHING

Now that we've got all the way to element 118 and completed the table, could we go further? The honest answer is that we aren't sure. Oganesson represents the filled seventh shell, but there might be an eighth or even a ninth shell.

Seaborg suspected the periodic table might stop when we reach element 126 because it's a magic number and beyond that the proton-repulsions may become too powerful, no matter how many neutrons we include. It has even been called unbihexium as a placeholder name.[20]

Other physicists speculate that we could go on to create a ninth period or a tenth and an eleventh without limit. We don't know enough about the nucleus to say for sure, so the only sensible thing to do is try. And that's the whole point of science: to see what might be possible.

CHAPTER NINE

Leftists

THE EASIEST NOBEL PRIZE EVER EARNED

Twenty years ago, if you'd asked a scientist what the most electrically conducting element was, they would have said silver. The only reason we don't use it in electronics is that copper is cheaper.

Then in 2004 two physicists won a Nobel Prize by making another element conduct better with a piece of Scotch tape.

Russian physicists Kostya Novoselov and Andre Geim (who you might remember as the scientist who made frogs levitate in 1997[1]) were working with graphite, the soft form of carbon used to make pencil cores. Because graphite is a brittle material it tends to become flaky and the scientists down the hall were using Scotch tape to clean their samples. By sticking tape to the graphite and peeling off the excess dust, the result was a shiny new surface.[2] It was watching this that gave Novoselov and Geim an idea.

If you stick the tape to a lump of already cleaned graphite you can extract a single layer of carbon, no more than one atom thick. This peculiar substance, which they named graphene, is arranged like a chicken-wire fence of carbon atoms and it has many unusual properties. Not only is it two hundred times stronger than steel, it is also transparent and can be used as a sieve to filter the salt out of seawater.[3]

On top of all this, graphene has an electrical conductivity better than silver. We measure electrical conductivity in units called siemens (pronounced zeemuns) per meter. Silver has a conductivity of 60 million siemens per meter and graphene clocks in even faster, although nobody has been able to agree on a definitive reading yet.[4] What makes this surprising is that carbon isn't a metal and it's usually only metals that conduct. Something very weird is going on.

WHAT IS A METAL?

When we hear the word metal we all picture the same thing: Ian "Lemmy" Kilmister, the bassist/vocalist of English rock band Motörhead. May he rest in peace.

After that we tend to think of grayish solids that are hard and shiny. What we're really thinking of when we do so are steel, titanium, aluminum, and chromium, the four metals that dominate our everyday experience, but metals have all sorts of other appearances and properties.

Bismuth forms labyrinthine square crystals, which glisten like oil on a puddle, while lutetium and thulium are found in fibrous clumps that look like pieces of torn beef. Niobium is a dull silver when first isolated but pass an electric current through it and it becomes rainbow-colored.

Some metals show magnetism (iron, cobalt, nickel, terbium, and gadolinium) while some are not magnetic themselves, but will reinforce the property in those five (neodymium). Some metals will remain solid when heated to over 3,000°C (tungsten) while others will melt in the palm of your hand (gallium). Their reactivity also ranges from gold, which won't even corrode in acid, to erbium, which explodes if you warm it gently.

With such a broad spectrum of behavior, what is it that unites them all? The answer is that a metal is an element that will always conduct electricity. Sure, carbon will conduct in the graphene state but metals will conduct no matter what state they're in.

In order to understand metal chemistry, we need to understand electricity and that story starts in ancient Egypt.

THE FIRST PHARAOH

In 3100 BCE the kingdom of Egypt was united for the first time under the rule of Narmer, the original pharaoh. There's a lot of debate around Narmer's true identity but we know the meaning of his name with some confidence. Narmer, translated into English, means "angry catfish."[5]

It may seem odd that a pharaoh would adopt the name of a river fish, but in Egyptian culture catfishes were the lord-protectors of the Nile and one of the most revered creatures in the world.

It's true that most catfish are useless monstrosities but the breed found in Egypt is special. Its Latin name is *Malapterurus electricus*, which means "electric catfish."

Like the electric eel of South America, this creature harbors a special organ that gives it the ability to deliver 400-volt shocks to anyone touching its skin. Records of the electric catfish are the earliest examples we have of electricity and it was five thousand years before humans could boast a similar control of the phenomenon.

SHOCKING

It is a crying tragedy that the man who discovered electricity is usually forgotten. The Greek scientist Thales (the one who fell down the pit) had already made the discovery that rubbing pieces of amber with wool caused them to gain a crackly property, which sparked under the right circumstances, but the discovery of what we think of as electric current goes to an English experimenter named Stephen Gray.

One of the reasons why Gray's work was overlooked is that he made the mistake of asking another scientist to help him develop it. That scientist was John Flamsteed, who happened to be a mortal enemy of Sir Isaac Newton.

Newton was a socially cruel, even malicious character who used his position as head of the Royal Society to discredit and bury the work of people he disliked, including Flamsteed.[6] Consequently, much of Flamsteed and Gray's achievements were ignored. It has to be said that while Newton was one of the greatest minds in history, he was also a jackass sometimes. So, let's redress the balance and give Stephen Gray his due.

Born in 1666, Gray worked as a dyer for most of his life and only indulged science as a hobby. He discovered electricity one night in his bedroom at the age of forty-two while playing with a crude instrument used to generate static—a tube of glass.

Static generators had been around since 1661, invented by the German politician Otto von Guericke, but Gray didn't have the money for such lavish equipment. He had to make do with rubbing a glass rod on rabbit fur and tapping it on whatever was around in the hopes of creating a shock.

Gray was curious about the fact that if you put the rod on the ground after rubbing it, it seemed to lose its electricity and wouldn't shock anything again until recharged.

On this particular night he decided to jam the end of the rod into a piece of cork and discovered that when he tapped the cork-tip against a pile of feathers, it sparked. The glass had been rubbed but the cork was somehow able to transfer the electricity through itself. Whatever electricity was, it could flow.

Excited by this result, Gray built a silk harness from his ceiling so objects wouldn't touch the ground, and began testing things to see if they would transfer electricity. After trying vegetables, string, coins, and anything else he could find, Gray began dividing everything into two categories: insulators, which wouldn't transfer electricity, and conductors, which would.[7]

The best conductors turned out to be metals, located on the left side of the periodic table. These were so good at electric transfer that Gray was able to pass a shock down nearly 250 meters of wire suspended from his bedroom window.[8]

Metals even conducted when pointing upward, which meant that whatever electricity was, it wasn't influenced by gravity. Electricity would still go into the ground, of course, but it's obviously not because of gravitational attraction. Instead, the planet itself was a conductor, which electricity will flow through given the chance.

Even more surprising among Gray's results was the fact that humans conducted electricity. By suspending a young boy from his silk harness, Gray was able to charge him and generate sparks from his face. This became the basis of a popular sideshow exhibit called "The Flying Boy" in which spectators could tap the floating youngster's fingertips and receive a shock.[9] All in the name of science.

The secret to this exhibition is that human skin is usually coated in a fine layer of saltwater in the form of sweat, allowing electricity to zap across its surface. When the spectators, who were connected to the ground, touched the charged boy, the electricity would flow over their skin and into the earth, creating the shock effect.

We know from Chapter 3 that electricity is made from electrons, so to explain all these behaviors we must turn once again to the Schrödinger equation.

STATIC

As we know, electrons occupy orbitals around their nucleus and atoms brought together can mix orbitals to form molecules.

Static electricity happens because this orbital mixing is not a rare occurrence. In fact, it happens when any two surfaces meet. As you sit on your chair right now, a few of the chair's electrons are forming temporary bonds with the electrons in your clothes (at least I hope so; please don't read my book naked).

When you stand up, most of the electrons return to their original atoms and the bonds are severed. Chair electrons go back to the chair and clothes electrons return to you. We refer to this as the triboelectric effect and it's a weak form of chemical bonding.

The thing is that some molecules are better at holding electrons than others and when it's time for the bonds to break they don't always return to their original configuration.

The molecules that make up human hair, for instance, are poor electron-holders whereas rubber is very good at it. If you put a piece of rubber such as balloon against your hair some of your hair electrons realize they're happier sticking with the rubber and they transfer across.

There's no limit to how many electrons you can cram onto a molecule so the rubber is happy to accept these travelers. When you separate from the rubber, some of your hair electrons stay on the surface of their new home and a charge imbalance arises.

The rubber and hair originally had no overall charge because the electrons and protons canceled each other but, if we transfer electrons from hair to rubber, things look different. The balloon finds itself holding an electron surplus while your hair has an electron deficit.

The surprising thing is that transferring electrons this way leads to greater stability. It sounds wrong because the rubber has stolen something from your hair, but remember that stability in quantum terms means "things have already lost energy to get to this state." Two molecules can be a lot more stable if they split, the same way a house of cards is a lot happier falling to pieces.

The overall result is that when you rub a balloon on your hair it steals somewhere in the region of two hundred billion electrons. That sounds like a lot but it's less than a trillionth of a percent of the electrons your body has in total.

If the balloon is now brought near a good conductor (like a piece of metal or the ground), the electrons are offered an even better deal and will flow into it, spreading as far away from each other as they can. Except this time we're not talking about little bonds being formed, we're talking about all the electrons jumping at the same time, creating the infamous static shock.

When Stephen Gray rubbed the glass rod, he was depositing electrons on its surface from the rabbit fur. Glass is an insulator so it can store electrons on its surface and won't allow them to flow from one end to another. The cork was a conductor, however, so the electrons were able to travel through it and into the ground. His experiments with wires were an extension of the same principle.

WHY DO METALS CONDUCT AT ALL?

As you read from left to right across the periodic table you're gradually increasing the number of protons in the nucleus. The more proton charge you have, the more electrons will be pulled inward and the smaller your atom becomes, meaning we see a decrease in atom size along each row.

Atoms on the left are therefore big and diffuse with great, floppy orbitals. Their electrons are also a long way from the nucleus with nothing much keeping them in place. This makes them ideal for sharing electrons with other atoms since the electrons have very little incentive to stay put.

When you get these bulky atoms together, their orbitals start mixing not just on a one-to-one basis but over the entire population. The atoms are so happy to share that when you solve the Schrödinger equation to describe millions of metal atoms, the result is a kind of mega-orbital—a turbulent free-for-all, which physicists call "the electron sea." This network of overlapping orbitals means electrons can easily slosh from one side of the structure to the other.

Touch any piece of metal and beneath your fingertips you've got a swarm of electrons flitting back and forth at will. These movements are random but if we can persuade the electrons to travel in one direction at the same time we have an electric current.

In smaller molecules, formed by elements on the right, gaps between the orbitals make it hard for electrons to move, so they won't conduct. That doesn't mean, of course, that it's impossible to force an electron through an insulator. Teflon, the most insulating material on Earth, can still be made to conduct but you need a fierce amount of energy to persuade the electrons to hop across the orbital gaps.

A substance with a conductance over 1 million siemens per meter is classified as a conductor while a substance below 0.01 is an insulator. Admittedly there's a huge gap between 0.01 and 1 million siemens per meter, but very few substances fall in this region. Those that do are deemed "semi-conductors."

THE WEIRDO

Whether a substance is a solid, liquid, or gas depends on how much the particles are attracted to each other. Oxygen molecules have little interaction because they're stable, making oxygen a gas at room temperature. It can be turned into a liquid by cooling it down (fun fact: liquid oxygen is blue) but under standard conditions it tends to spread out.

By contrast, metals are good electron sharers, meaning their orbitals overlap and they clump together forming a solid, with the obvious exception of mercury, the metallic liquid. A full explanation for mercury's liquidity requires knowledge of Einstein's theory of special relativity, but we can get the gist without worrying about that.

Like other metals, mercury's orbitals stick out in many directions like petals on a flower so it can conduct, but it's in a funny position on the table. It sits on the bottom row, making it huge, but over on the right-hand side meaning it has a lot of protons pulling the orbitals inward. The result is that the orbitals are extended enough to overlap but not quite enough to hold the atoms together.

Move to the right and you increase the proton number, causing the atoms to pack together better, resulting in a solid. Move to the left and the orbitals overlap better, also resulting in a solid.

Mercury atoms are just too weakly attracted to stick together, but just attracted enough to allow electrons to hop from atom to atom. The result is that mercury is a conducting element and therefore a metal but it's unquestionably the worst metal on the table.

THE OTHER WEIRDO

When electrons travel through a piece of metal they don't move in perfect lines. The nuclei vibrate and the inner orbitals interfere with the outer ones. The result is that conductivity never happens perfectly and we call the collection of things that slow it down "resistance," measured in ohms. The energy electrons are given as they are pushed through the metal is called voltage (measured in volts). These things together give rise to the overall electron flow.

If we think of voltage as a fist squeezing the end of a toothpaste tube, the resistance is the diameter of the tube and the actual amount of toothpaste that comes splurging out is what we call current, measured in amperes (amps for short).

A watch battery delivers electrons into the watch with an energy of about 1.5 volts. The resistance of the circuit slows it down and we end up with a current in the region of five-millionths of an amp (0.000005 A).

For perspective, a bolt of lightning packs around 100 million volts. This electricity is forced through the air, however, and the overall current ends up at around 5,000 amps by the time it reaches the ground. Passing electricity through non-metals like air involves a lot of energy being lost.

Graphene's conductivity is therefore very strange. Carbon is a nonmetal most of the time but when it is arranged in the thin wafers of graphene it starts to conduct.

It happens because the atoms in graphene are arranged in flat hexagons with each atom bonded to three others. Since carbon has an available four electrons in its outer orbitals, each atom has a spare one that isn't involved in bonding. This electron can move from atom to atom with hardly any obstruction, so even a small voltage will produce a lot of current.

Where graphene differs from metals is that it is almost two-dimensional. In a metal, electrons can change route and go exploring in all directions but in graphene there are less places to go. It is practically a flat plane, meaning electrons have no possibility of moving up or down, making them more likely to stay on track.

ELECTRICITY AND YOU

In 1886 an American human rights committee decided that execution of criminals by hanging was inhumane and a new method of capital punishment was needed. One of the people on the committee was Alfred Southwick, a dentist from New York who had already designed an electrocuting chair several years previously. Southwick's idea was approved and testing began, endorsed by none other than Thomas Edison himself.[10]

At the time, there was a battle going on over which type of electricity the United States should adopt. Edison had put a lot of money into battery-based electricity and needed to find a way of tarnishing the reputation of magnetically generated electricity, favored by his rival George Westinghouse. His solution was simple, if a little gruesome.

In the most morbid marketing strategy ever employed, Edison insisted that the newly designed electric chair be configured to run on Westinghouse's electricity, so people would associate it with death.

He tested the chair on stray animals in his workshop and is on record as having killed dogs, cats, birds, a horse, and a circus elephant named Topsy (he was considerate enough to film that last one, and you can watch it for free online if you're into that sort of thing).[11]

Soon after, the electric chair was rigged-up for its first victim, William Kemmler, in 1890.[12] Kemmler took over four minutes of continual electrocution to die, with the procedure stopping halfway through until someone screamed, "Great God, he is alive!"[13] Humane indeed.

The key to the electric chair is making sure the human body is part of a circuit, which is actually quite difficult to do. Despite what Saturday morning cartoons claim, you're not very easy to electrocute.

If you're ever unfortunate enough to be hanging from a power line you may feel a tingling in your fingers, but you're in no real danger. Once the electricity has filled all the available orbitals on your surface there is nothing else it can do.

If, on the other hand, you somehow connect to the ground then you're not a cul-de-sac anymore: you're a pathway and the electricity will use it. If electricity goes onto you, you're fine, but if electricity goes through you, you're in trouble.

The human body is a fairly decent conductor (you're a bag of salty water) but to complicate matters your skin is an excellent insulator. Dry skin has a resistance of about 100,000 ohms, although wet skin absorbs water into its pores and the resistance drops to around 1,000 ohms.

It's also worth pointing out that once electricity enters your body it will travel along the easiest path available. A tiny amount may go exploring, but you could pass thousands of amps through your hand without dying. It would still hurt, so don't do it, but you wouldn't be in mortal danger.

The only time electricity becomes lethal is if it passes through your heart, lungs, or brain for a sustained period of time.

The way your heart works is that the muscular outer layer is given a short electric shock of around 0.0000012 amps every second, which causes it to contract, squeezing blood to your body. Afterward it is allowed to relax and reopen, taking in more blood, before the whole thing is repeated.

If a current is pushed through the heart for a long time, however, it squeezes tight and doesn't reopen, meaning it can't take in a fresh load of blood. That's why people can survive lightning strikes but not the electric chair. The electricity of lightning may pass through your heart, but it does so for a short time only and your heart is able to return to normal. If you keep the current flowing, you essentially give the person an artificial heart attack.

Surprisingly (or perhaps not) very little research has been done on how much current is needed to make the heart do this. The approximate guidance, based largely on anecdotal evidence and a bit of bioelectrical theory, suggests that around 0.05 amps is required to kill a person.

The electric chair worked by passing a current of between 1 and 7 amps through the body depending on the state legislation. That's over twenty times the lethal dosage.

Typically, the two live ends of the circuit would be connected to the scalp and ankle so the current would pass through the brain, heart, and lungs together, guaranteeing the malfunction of at least one of them and ensuring a warm death. Have a nice day.

CHAPTER TEN

Acids, Crystals, and Light

A BARREL OF HORRORS

In March 1949, English newspapers reported one of the most gruesome crimes to occur in British history since those of Jack the Ripper. John George Haigh, who the *Daily Mirror* had referred to as the Vampire Murderer on March 3, was taken to court and charged with six counts of pre-meditated homicide. What made them particularly ooky wasn't the murders themselves but the way he disposed of the bodies.

After drinking glasses of their blood, Haigh loaded each body into a 40-gallon drum, which he topped with concentrated sulfuric acid and left for two days. The remaining sludge was poured into a drain behind his workshop, earning him his other colorful nickname, the Acid Bath Murderer.

Acids capture people's imagination because they are the standard "nasty" chemicals, capable of chewing through a human body and destroying any evidence the person was ever there. The only reason Haigh was caught was because the solution of his final victim, Olive Durand-Deacon, still contained part of her plastic denture, which her dentist identified.

Haigh was executed by hanging on August 10, 1949. He claimed to have other victims but they are unidentified to this day because the bodies were disposed of so perfectly.[1]

IT BURNS, IT BURNS!

An acid is a substance whose molecules fall apart in water to produce free-floating protons. Protons are the charged particles inside a nucleus, shielded most of the time by their electron orbitals, but if they get released when their parent molecule dissolves they can cause untold damage.

A rogue proton is a concentrated lump of charge and will pull electrons toward itself at any cost. Things like glass or plastic have strong bonds between atoms so acids aren't usually able to react with them, but any chemical with loose bonds, including the ones in your body, will be pulled apart.

An acid can be thought of as a proton juice and the easiest way to generate a solution of protons is to make sure your starting molecule contains hydrogen. Hydrogen is the simplest element, consisting of one proton and one electron, so if its electron is more interested in the other atoms of the parent molecule, the proton will drift away.

Take hydrogen chloride. Each molecule consists of one H atom and one Cl atom, giving it the formula HCl. The chlorine atom is very good at holding electrons, better than the hydrogen, so the bond between them isn't a fifty/fifty share—it's lopsided like so:

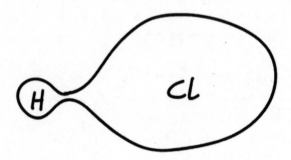

Put the whole thing in water and the two atoms will separate with chlorine keeping all the electrons and hydrogen being left essentially naked.

This lonely hydrogen proton drifts away, waiting until some other molecule with which it can react comes along. We have generated hydrochloric acid, the one in your stomach, capable of dissolving bone.

THE STRONGEST ACID

We measure how strong an acid is by how willing it is to let go of a proton. The numbers involved spread over an enormous range and we use something called the pK_a scale to measure it. The scale works the same way as the earthquake Richter scale where each number is ten times greater than the one before it. The scale also works backward for reasons we don't need to worry about (see Appendix V if you're curious). So, the lower the number, the stronger the acid.

Household vinegar has a pK_a of 5 whereas oxalic acid, the one in rhubarb, is closer to 4, making it ten times more potent. Then there's chromic acid, a powerful industrial agent, with a pK_a of 1—three places lower down the scale than oxalic and therefore one thousand times stronger. For context, you can eat oxalic acid and feel fine but chromic acid will set fire to living tissue.

The concentrated sulfuric acid Haigh used to dispose of his victims scores a –3 on the pK_a scale, seven numbers lower than vinegar and therefore ten million times stronger.[2] That's another way of saying that sulfuric acid is ten million times better at releasing its proton than vinegar. But if we can create molecules with absolutely no interest in holding their hydrogen, we end up with a class of chemicals on our hands called superacids (well, hopefully not on our actual hands).

Perchloric acid has a pK_a of –10, which is ten million times stronger than concentrated sulfuric and triflic acid has a pK_a of –14, one hundred billion times stronger.[3] And that's not even touching on magic acid (actual name), which will dissolve even candle wax.[4]

If you scan around the internet, most popular science websites tend to report the strongest acid in the world as something called fluoroantimonic acid, boasting a pK_a of –19. That's ten quadrillion times stronger than sulfuric. It's occasionally used in the electronics industry to etch equipment, but it doesn't really deserve the gold medal. That goes to an acid so strong it has only been synthesized once in recorded history.[5]

An acid's job is to kick hydrogen away so the best acid will be one where the other atoms don't want to bond with it in the first place. And there's no better atom for that than helium, the least reactive element on the table. If you can force hydrogen to bind with helium, you've created the weakest bond it's possible to get and it will fall apart instantly.

In 1925, the chemist Thorfin Hogness managed successfully to brew a microscopic quantity of helium hydride that possesses a pK_a of, brace yourself, −69.[6] That's so strong there isn't even a word to describe how much better it is than sulfuric acid.

The non-reactivity of helium is also responsible for another record-breaking property it possesses: liquid helium is the most fluid liquid in the Universe. When a sample of helium is cooled to around −269°C, the atoms lose their movement energy and settle into liquid form. In most liquids the atoms still interact with each other a little, but in helium they keep to themselves.

If you take a cup of liquid helium and stir it once, it will keep spinning forever. Any other liquid would interact with the container and be slowed down but liquid helium doesn't feel friction and will keep spinning until the end of time.[7]

Wouldn't that constitute a perpetual motion machine, though? The answer is that if we tried to put something like a propeller into the swirling vortex, the helium would just flow around it. The only way to get the liquid helium to work on something would be to warm it up, and as soon as you do that the superfluidity is lost.

Liquid helium also happens to defy gravity. Air pushes down on everything at atmospheric level and near the edges of a container some liquids are able to creep up the sides because they're being pushed by air on one side but not the other.

Most liquids are self-attracting enough to stay together and not begin climbing the walls but liquid helium isn't most liquids. Helium will move up the sides of an open container and creep its way out, emptying the vessel as if it had a desire to escape.

In order to understand the surreal properties of liquid helium and helium hydride, we're going to travel to the right side of the periodic table. The realm of the non-metals.

SELFISH CREATURES

Most of the spectacular and violent reactions in chemistry take place in the non-metals because they're so greedy. As we've already seen, metals are big and have friendly, overlapping orbitals, but atoms on the right are small and grip their electrons tightly.

The most reactive element is fluorine, which we met in Chapter 1 when it was setting fire to water. A sparse yellow gas, fluorine needs to be transported in dense steel and bulletproof glass because it will rip the electrons out of anything else it touches.

Because it's so electron-hungry, a molecule of two fluorine atoms will be perfectly symmetrical as the electrons are shared between them. If you bond it with a metal such as cesium, however, the bond is uneven, with fluorine getting the lion's share of electron density. It's similar to the way hydrochloric acid molecules are arranged—non-metals always win because they don't like to share.

This electronic exchange means cesium atoms become electron deficient while the fluorine atoms become electron rich. It's not really correct to call them atoms anymore since they aren't neutral units so we refer to them as "ions" instead.

Ions are still sharing electrons but it's such an uneven bond we usually just imagine cesium losing electrons and fluorine gaining them.

You'll see diagrams of ionic bonds in which the particles are drawn like balls packed together, such as in the diagram on page 122 (top). That's not strictly correct, but it helps keep track of where the ions are and how they're arranged. The diagram at the bottom gives a slightly more accurate picture.

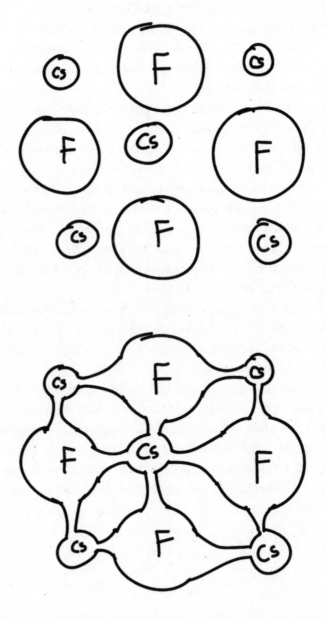

This type of bonding where things are arranged in a lattice frame-work gives rise to crystal properties. Starting at group 13 with boron, non-metals tend to keep their electrons in very specific places, forming grids with sharp edges. This brings us to . . .

SPARKLY CREATURES

Boron is the second-hardest element after carbon. Used in making glues and glass, it's usually found bonded to oxygen and sodium in the form of borax crystals exported from Death Valley, California, the hottest place on earth.

Borax crystals have a ghostly white appearance and are mostly transparent, something you don't get with metals. Because of the sea of electrons, light will bounce off the surface of a metal, making it opaque. Non-metals, on the other hand, hold their electrons in fixed orbitals with gaps, meaning beams of light travel through rather than getting reflected.

Depending on the angles between ions and the sizes of their orbitals, a beam of light can emerge from a non-metal looking very different to when it entered. As the light is bounced around inside the crystal matrix, it can lose or gain energy, changing color and giving the crystal a different appearance.

The most common crystals on Earth are based on silicon and oxygen in the form of SiO_2. It's the other elements mixed with them that give rise to the different minerals we find in the ground. A single hunk of rock (a conglomerate of mineral crystals packed together) can contain dozens of different elements and we have to extract them with acids or electricity.

In fact, most elements on the table were discovered by grinding up rocks and seeing what was inside them. The elements yttrium, ytterbium, erbium, and terbium, for instance, were all discovered at the same Swedish mine from a single type of rock.

The most prized crystals, though, tend to be based on oxygen bonded to aluminum rather than silicon. On its own, aluminum oxide is a white crystal called corundum with an appearance similar to table salt. But if a few chromium atoms get mixed in, you've got ruby. Replace the chromium with titanium or iron and you get sapphire.

Then the most precious crystals of all, diamonds, are made from carbon atoms forming a tetrahedral array, with each atom linked to four around it. And again, it's the impurities that give the colors. A bit of boron and your diamond turns blue while a touch of nitrogen will give you yellow. Change the atoms and you change the color.

HIGHFALUTIN ELEMENTS

As we go across any row on the periodic table we're dealing with atoms that house more and more protons. The electron orbitals get sucked in and, as a result, everything on the right is smaller and greedier.

Group 17 is where we get things like fluorine (sets fire to cotton wool), chlorine (a chemical weapon), and bromine (a toxic disinfectant). But when we get to group 18 something strange happens. The elements of this column—helium, neon, argon, krypton, xenon, and radon—are the least reactive on the table.

They are so reluctant to get involved with bonding that they were originally named inert gases. We have since learned that group 18 elements will mingle with others a little, but not if they can help it. As a result, these snooty substances are referred to as "noble" gases (other groups have names too, see Appendix VI).

We saw earlier that helium's refusal to bond is what makes helium hydride the strongest acid in the world, so the obvious question is what are the most lifeless elements doing next to ones like fluorine and chlorine?

The answer comes from how electrons are distributed around the nucleus. Orbitals are fixed in certain shapes according to the quantum rulebook but they are also grouped at specific distances.

The first set of orbitals are huddled around the nucleus, but the second set are a great distance away. This outer set is repelled by the inner set and there is a no-man's-land between them.

The diagram opposite shows the energy levels of the first and second orbital sets. For simplicity, we're ignoring the orbital shapes because then our diagram would look like a plate of panda entrails.

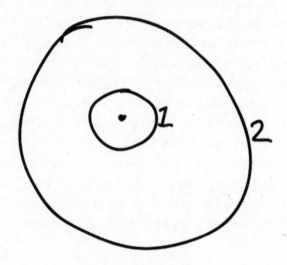

We call these orbital groups "shells" and they are the reason for the periodic trend Newlands identified. As we go from one side of a row to the other, we are filling the orbitals of a particular shell. When a shell is full, we jump to a higher one and start filling that instead, starting a new line on the table.

The noble gases are the elements we get when we have completely filled a shell. Because every orbital of these atoms is full, there's nowhere to put an incoming electron. The atoms are also small (they're on the right-hand side), which means they hold their own electrons tightly and won't donate to anything else.

Noble gases are therefore unlikely to accept electrons or donate them, making them bad at bonding. A few dozen noble-gas compounds have been created over the last few decades but it's not a common occurrence.

It might seem that these elements are pointless and boring, but their refusal to react makes them useful. Take a light bulb. The filament inside is made from tungsten, which glows when electrified. The problem is that the tungsten gets so hot it would begin reacting with oxygen. To avoid this problem, we flood lightbulbs with argon instead of air so nothing reacts and the bulb can continue to do its thing.

We can also use noble gases to produce vibrant colors of their own. If you trap a sample of noble gas inside a glass tube and pass a current from one end to the other, the atoms will start vibrating. The electrons get pushed outward by the electrical energy, but they fall back immediately, kicking out specific beams of light.

Any other gases would start reacting and everything would rearrange to become stable. Since stable means no more energy is available, your light would switch off moments after you switched it on. But noble gases are so reluctant to bond they just keep jumping back and forth, emitting a steady stream of light. Neon makes the tube glow red, helium glows orange, argon glows blue, krypton glows green, and xenon glows turquoise. Neon was the first to be discovered so we refer to these harsh buzzing tubes of gas as "neon lights," the kind you see outside store windows.

CHAPTER ELEVEN

It's Alive, It's Alive!

THE MOST TOXIC POISON

In 2006, the world's media reported on the agonizing death of Alexander Litvinenko as he succumbed to polonium poisoning. What made the story so chilling, aside from the political overtones, was the minuscule amount of polonium needed to cause death. It was estimated that Litvinenko consumed less than one-hundredth of a gram and was dead within three weeks.[1] Is polonium the worst thing you can have in your body?

Judging toxicity is not as straightforward as you might imagine. For starters, everyone metabolizes things differently. Nicotine alone turns into seven different chemicals depending on the person, which might explain why some people find it harder to quit smoking. They literally turn it into more addictive substances.

This means that if you poison a large group of people some will die and some will survive, purely by chance. In order to get around this, biologists use something called the LD_{50} value, the lethal dose guaranteed to kill 50 percent of a group. The number is given in mg/kg (how many milligrams needed to kill every kilogram of creature) and the lower the LD_{50}, the more toxic the substance.

The LD_{50} of pure caffeine is 367 mg/kg.[2] A baby duck, which typically weighs around 1 kg, could therefore ingest 367 mg of caffeine and have a 50 percent chance of living. An African bull elephant, on the other hand, weighs 5000 kg so you'd need about 2 kg of caffeine to be 50 percent confident of killing it.

It's also difficult to quote accurate LD_{50} values for humans because the only way of obtaining them would be to poison a bunch of people and see how many died. Sadly, there have been cases of experimentation on unwitting suspects, but usually such studies aren't common.[3]

Some animals can be seen as close approximations to humans but you run into the same problems. Different species metabolize things differently. Glucuronic acid is harmless to humans and used in cooking sauces but it's lethal to a cat. Arsenic is toxic to us but when added to chicken feed it causes them to gain muscle mass. Plus, there's the well-known fact that theobromine in chocolate can kill a small dog, but all it does to humans is leave them with a sense of self-loathing.

The animals biologically closest to us aside from chimpanzees, which are not tested on, are rats. Whatever your ethical stance on animal testing, the fact remains that trialing a chemical on rats is the closest thing we can get to human data.

It's also worth remembering that chemicals get processed differently depending on how they are absorbed. Some elements like holmium are toxic no matter how you take them, but something like indium is only dangerous if inhaled. (NB: probably best you don't ingest either.)

All of these factors make it very hard to say what is the most poisonous chemical in the world. That's probably a good thing, but since we're on the subject we might as well look at some of the candidates.

Lead has an LD_{50} of 600 mg/kg while thallium has one of 32 mg/kg, making it twenty times more dangerous. Arsenic, the preferred poison of nineteenth-century novelists, has an LD_{50} of 20 mg/kg while phosphorus comes in at close to 3 mg/kg.[4]

If we go on toxicity alone, this makes phosphorus the most poisonous element, but if we include the effects of radioactivity polonium outranks it by a clear mile. Radioactive elements don't just kill by interfering with the functioning of the body: they spit alpha particles (see Chapter 8), which essentially rip your cells apart.

Because of this additional mode of killing, polonium probably is the deadliest element. In fact, nobody knows what the LD_{50} of polonium is because experimenters are reluctant to work with it. Even its dust can kill you. But given the amount needed to kill Litvinenko, the LD_{50} is going to be very small.

If we start including compounds as well as elements, though, polonium is no longer that bad. Dimethylcadmium is often cited as the most toxic compound in the world, so toxic that a thousandth of a gram dissolved in a ton of water is lethal.[5] But the crown really belongs to botulinum toxin, a chemical produced by the bacteria *Clostridium botulinum.*

There are several varieties given the names A to H, and it's botulinum toxin H that is the worst. Only two-billionths of a gram are needed to kill a fully grown adult.[6] Assuming the population of Earth is around seven billion, you would therefore need only 14 g (a teaspoon's worth) to wipe out the entire species. And it kills you in a pretty nasty way, paralyzing you to death.

You can also dilute it down to low concentration and inject it into your forehead, paralyzing the muscles and preventing wrinkles. Botulinum toxin A (not quite as deadly) is used for exactly this purpose and is marketed under the trade-name of Botox®.[7]

THE ELEMENTS OF LIFE

In 1924, the head of the American Medical Association, Charles Mayo, published a tongue-in-cheek calculation showing that if you split a human body into piles of its constituent elements the total value would be around eighty-four cents.[8] The iron from your blood would make a single household nail while the carbon in your proteins would make a small bag of charcoal, etc.

We did something similar in the introduction when we looked at the chemical formula for a person. It's a powerful reminder that the atoms that make up our bodies are no different to the atoms that make up the contents of our kitchen.

A lot of people seem uncomfortable with this notion. I once saw a magazine ad in which a worried customer is reassured by a scientist that their ice cream "contains no 4-hydroxy-3-methoxybenzaldehyde, only natural vanillin." What the writers of the advertisement didn't seem to realize is that 4-hydroxy-3-methoxybenzaldehyde is just the chemical name *for* vanillin. It would be like saying "this drink contains absolutely no H_2O, only water."

During the Middle Ages, everyone thought living creatures were made from magical "essences" different to non-living things. It was a belief called vitalism, but like most ancient quackery the cracks were beginning to show by the Renaissance.

In 1745, Vincenzo Menghini burned human organs to ash and discovered that you could extract iron powder from the remains with a magnetized knife.[9] He concluded that humans had to contain the base metal iron and that perhaps we weren't made of magical ingredients after all.

In 1828, Friedrich Wöhler went even further by manufacturing urea from cheap lab chemicals.[10] Urea is the main component of urine and therefore was assumed to be beyond human understanding. Wöhler showed it to be a bog-standard molecule with the formula CH_4N_2O.

Whether you like it or not, the elements used in living biology are no different to those used in sterile chemistry. A strand of human DNA contains 204 billion atoms, all of them carbon, hydrogen, oxygen, nitrogen, or phosphorus. There's no additional "essence" to make it special.

The iron Menghini discovered is used in blood to bind oxygen molecules and transport them to the various organs. When the oxygen gets to where it's needed, enzymes and proteins containing chromium, molybdenum, copper, and zinc help store it, while manganese holds harmful atoms in place before they cause damage.

When a woman is pregnant she spends nine months breaking down food and reconstituting the atoms into a baby. The calcium in milk becomes the calcium in your bones, the nitrogen in potatoes becomes the nitrogen in your skin, and the sodium in salt becomes the sodium in your brain. In a very literal sense, we are what we eat.

It's not just animals either. Plants use magnesium to absorb sunlight and vanadium or molybdenum to bind nitrogen from the soil, a crucial nutrient in growth. It doesn't matter what the biological system is, you'll find every bit of it on the periodic table.

I've occasionally heard people referring to biology as applied chemistry because of this deep connection, but this isn't fair at all. Biology is just chemistry at its most wonderfully elaborate.

But it comes at a price. Since we are made from the same stuff as the world around us, that makes us vulnerable to the same malfunctions.

STRIKING A BALANCE

During the 1500s, Germany was going through a scientific renaissance and one of its most prominent figures was the great Swiss physician Paracelsus. His real name was Theophrastus Bombastus von Hohenheim and he was the first person to investigate medicine as a science rather than a superstition (although he did believe in gnomes—nobody's perfect).

His most famous dictum is named the Paracelsus principle in his honor and it's simple: "the dose makes the poison." In other words, whether something is beneficial or harmful is all about the quantity.

Even something like cyanide is only harmful above a certain level. In fact, apple seeds contain amygdalin, which your body converts to cyanide, but you'd need to eat the seeds from about eighteen apples in order to get sick (assuming radioactive bananas don't kill you first).

The metals in your body are the same. If you don't have enough copper then your immune system can't function, but if you get too much your eyes turn reddish-golden. Beautiful for sure, but you won't appreciate it since you'll be vomiting blood at the same time.

The element arsenic is famous for its use as a poison, but in small doses it can treat leukemia.[11] It was also the central atom in Salvarsan, the world's first wonder-drug and the main reason we don't hear much from syphilis these days.[12] Antimony can be administered as an anti-bacterial agent but too much starts killing the host, and a small amount of cerium can treat tuberculosis but too much gives you a heart attack.[13]

The Paracelsus principle is why your medication has a recommended dosage. Get the amount of chemical right and you save lives, get it wrong and you end them.

WHY ARE THINGS POISONOUS IN THE FIRST PLACE?

The honest-to-God truth is that we don't know why some things are bad for you and others are good. Given the number of chemical compounds that exist, it would be impossible to catalog the effect of each one. We've only known about molecular bonding since the late 1920s so it's no surprise that much of biology is still out of our reach. It's been doing its thing for over three billion years so there's no way we'll have it figured out in a century.

Humans are a delicate balance of reactions. If we alter one of them we can trigger a chain reaction and the final outcome can be unpredictable.

For example, if you get too much of the element tellurium in your body it causes horrendous breath and elemental silver will turn your skin blue, a condition known as argyria.[14] Even nitroglycerine, which we met as the active ingredient in dynamite, is used to treat angina and nobody's sure why it works.[15]

One of the few poisons the actions of which we do have a good understanding is cyanide. It works because cyanide molecules bond strongly to iron. If they happen to bond to the iron at the center of a molecule called cytochrome c oxidase, the iron can no longer be used and the whole thing shuts down.

This is bad news because cytochrome c oxidase is the molecule we need to extract energy from food. Switching it off means we essentially starve to death in a matter of minutes rather than weeks.

We also know that some elements, particularly heavy metals, are poisonous because they're similar to elements your body needs and enzymes can accidentally incorporate them.

Zinc is needed for growth, and the element cadmium has a similar size so if you ingest it the body starts building enzymes with cadmium instead. Cadmium doesn't have the right orbitals to interact with the chemicals in your body, however, and the result is that you suffer from cadmium poisoning. Your body stops growing.

Lead poisoning occurs because lead is a similar size to calcium, needed to manufacture red blood cells, so if your body absorbs too much lead you can't make blood. Mercury is even worse because it's the right size to fit through membranes surrounding your brain. Once it gets inside, it can affect your nervous system, not to mention your thought patterns.

Most people avoid mercury for this very reason but, during the nineteenth century, warm mercury nitrate was used as a key ingredient in preparing hat felt. Sure enough, people in the hat industry soon got a reputation for being a few electrons short of an atom, hence the term "mad as a hatter."[16]

THE FIRE WITHIN

Running all these reactions is exhausting for your body, requiring a constant supply of energy to stay alive, so you obtain it by taking in sugar and setting fire to it.

In a chemical context, sugar doesn't refer to one chemical but a collection of them. They're all made from carbon, oxygen, and hydrogen atoms looped into hexagons or pentagons, and the stuff in your kitchen is a mixture of two kinds called sucrose and fructose. The different types of sugar you can buy, such as granulated, powdered, icing, etc., refer to the size of crystals rather than the chemicals themselves.

Most of the food we eat contains sugars, which the body breaks into the smallest type, glucose ($C_6H_{12}O_6$). The glucose molecules then enter a sequence of reactions that convert them into water and carbon dioxide. The water is lost through sweat and the carbon dioxide dissolves into your blood where it is carried to the lungs and breathed out. The air you're exhaling now is made from the food you ate this morning.

The original C, H, and O atoms are then repackaged into a highly unstable molecule called adenosine triphosphate, or ATP for short. ATP has a chain of phosphorus oxides hanging off it, which will detach at any given moment, releasing light and heat as they do. This energy can be absorbed by other molecules and gets used to drive all the reactions in a cell.

The whole procedure is controlled by molecular machinery swinging in and out to ensure the correct reactions happen at the correct time, and its discoverer, Hans Krebs, scooped the Nobel Prize for mapping the whole carnival.

This is the reason we need food in the first place. Without sugars we couldn't supply energy to drive all the other chemical reactions that make us a living thing. With the exception of one species (*Spinoloricus cinziae*, which seems to have evolved a different way of getting energy), every creature on Earth carries out Krebs's reaction.

It's called respiration from the Latin *spirare* (to breathe) and it's the same thing, chemically speaking, as fire. Some chemical reacts with oxygen, producing carbon dioxide, water, heat, and light in the process. We are all walking fire factories.

The only reason we're not in danger is because it happens in several stages and on a very small scale. Just as well, because otherwise we'd burst into flames spontaneously. Speaking of which . . .

THE FURNACE WITHIN

The earliest record of spontaneous human combustion was the death of an unnamed Polish knight during the early sixteenth century under the reign of Queen Bona Sforza. The account appears in a 1654 book written by Thomas Bartholin who heard it as a secondhand account from Adolphus Vorstius, who heard it from his father, who claimed he once saw a document where it was written.[17] Originally in Latin, the brief description translates as "he drank two cups of warm wine, then belched flames and was toasted."

Spontaneous human combustion (SHC) is a controversial subject because nobody agrees on whether it happens. The idea that a person can catch fire without external ignition is very dramatic but apparently so rare that solid research is impossible to find. It's not like you can study a group of people and see which one of them spontaneously combusts. It's spontaneous.

Most SHC reports are like those of the Polish knight above; spurious secondhand descriptions and probably nothing more than ghost stories. Plus, the accounts that do give details are usually easy to explain. But it's an interesting topic that captures the imagination so it's worth looking into.

In most cases of SHC, the remains of a human body are found charred or melted with the exception of the feet and hands. The bones are turned to ash and, in most instances, the surrounding furniture is untouched.

Let's address the bones turning to ash first. Many people argue that the temperature of such fires must be fierce in order to have such an effect. After all, the furnaces in crematoriums typically run to over 980°C.

However, the need for these high temperatures is because crematoriums have to burn a body quickly. A flame of a few hundred degrees is still enough to turn bones to ash provided it's left for several hours. If you have a fuel source that lasts that long, there's no mystery.

Explaining the fuel is the next task and in 1998 a scientist named John de Haan conducted a series of experiments in which he wrapped a pig carcass in cloth and set fire to one corner. Once the ignition had been provided, the water content of the pig boiled away and the dry carcass continued burning for five hours, destroying everything apart from the trotters.[18] The explanation for this gruesome demonstration is "the wick effect."

The subcutaneous fat of most mammals is flammable so, if the skin is broken, it can melt and leak into the surrounding cloth. The fabric is now doused in liquid fat and will burn like a candle wick for hours, using the full supply of body fat as fuel. This also explains why feet and hands are the only things left over; they have very little fat content so the fire leaves them unscathed.

So how come the rest of the room is always left alone? We're used to hearing about fires getting out of control and buildings burning to the ground because fire will supposedly spread and destroy everything in its path. But if we really think about it, we know that isn't true.

Most fires stay put and combust upward, not outward. Unless the ceiling is very low a fire will usually have nothing else to burn once the fuel is exhausted. Think of how you're able to stand right beside a bonfire or hold a flaming match without your skin catching fire. Or think of all the exercise books sitting in chemistry labs the world over, inches away from Bunsen burners, none of them burning.

You can hold a piece of tissue paper an inch from a flame and it still won't catch. Even if you waft it through the fire itself, it will only warm up.

Fires that do spread and make the news are usually the result of direct contact. A forest fire proliferates because the trees are touching each other or the wind is blowing flames from one place to the next. Contrary to gut feeling, fires do not spread through air with ease—otherwise we'd set the atmosphere ablaze every time we switched on an oven or lit a cigarette.

Provided there is something to start off the fire, the discovery of an SHC victim is not suspicious at all and actually goes along with straightforward science. And it turns out that in most detailed reports of SHC there is an obvious source of ignition.

For example, the death of Nicole Millet (February 20, 1725, Rheims, France) is often cited as spontaneous human combustion since she was found on the floor charred to a crisp with little damage to the surroundings. What has to be factored in is that Millet was a heavy drinker and had gone to "warm herself by the fire" with a bottle of alcohol.[19] Hmm.

Similarly, Mary Reeser (July 2, 1951, St. Petersburg, Florida) was found torched in an armchair, again with little damage to the room apart from the chair she was sitting on.[20] After investigation, however, the FBI concluded that Reeser was taking sleeping pills, which caused her to fall asleep while smoking.[21] Hmm.

As scientists, we have to be skeptical, particularly of strange claims. In most cases, it turns out that while human combustion can happen there is nothing spontaneous about it. And yet . . .

I don't know whether spontaneous human combustion happens. Almost all the claims turn out to have obvious causes, but I cannot ignore the fact that one or two do not. Of the few hundred documented cases of SHC in history, there are a handful that seem to defy explanation.

The case of Robert Francis Bailey (September 13, 1967, Lambeth, London) is one such instance. A group of people walking outside a vacant house in London reported a bright flickering light inside and called the fire department, who arrived within minutes. When they entered the house, Brigade Commander John Stacey reported, Bailey's body was curled on the floor with a four-inch slit in his stomach from which a roaring flame emanated. The house's electric and gas supply had been disconnected and there was no sign of matches anywhere.[22] So how did the fire start and why was it bursting from his gut?

Then there's the account of Raymond Reed, who was with the Ninth Battalion of the Royal Welsh Fusiliers during the Second World War.

Reed didn't combust himself, but recounts one night in Dorset when he was crossing a field and a nearby sheep exploded.[23] Presumably the sheep wasn't smoking in bed.

There's also the 1867 case of Mr. Watt of Garston whose corpse suddenly began to burn in a church crypt long after his death from typhoid fever.[24] Not only is it unlikely he was smoking in bed, he was encased in a coffin.

Accounts like these, if they are to be believed (and that's a big if), are difficult to rationalize. The wick effect would explain the remains, but there doesn't seem to be a source of ignition.

We must be careful, though. Just because we don't have an explanation for something doesn't mean we have to accept a fanciful one. These accounts can't be explained, but the sensible thing to do is say we don't know the explanation, not put in any hypothesis that we like. There's no reason to assume SHC unless we can find evidence for it directly. Otherwise, we might claim that every unexplained fire is the result of spontaneous combustion.

There is, however, one detail that pervades every account of witnessed spontaneous combustion and might just qualify as potential evidence. The flames are always reported to be bright blue and originating in the gut.

In 1993, Gunter Gassmann and Dieter Glindemann showed that the interior of the human gut is capable of forming a chemical called phosphane (PH_3).[25] By itself, phosphane isn't flammable, but if two phosphane molecules are linked together they form diphosphane (P_2H_4), which is. Diphosphane can spontaneously ignite in the presence of oxygen and burn the other gases in the vicinity. The main gas within the human body is methane (CH_4), mostly found in the gut and famous for its blue flame.

Diphosphane often forms in marshland conditions, which is why people occasionally report blue flames around swamps and graveyards. So-called will-o'-the-wisp ghosts are actually methane fires triggered by phosphorus chemistry.

At present, there is no known mechanism that causes diphosphane to form inside the intestines, but *if* there was and *if* it came into contact with oxygen and *if* there was enough methane present, there is a slim chance a fire could conceivably start.

The scientifically honest answer to whether spontaneous human combustion can occur is still "we don't know." Diphosphane offers a tantalizing possibility, but speculations are not proofs. What we can say is that if spontaneous human combustion really does happen, it's a one-in-a-billion chance.

I have made it clear to my friends that if I happen to be one of the few people who dies from spontaneous human combustion, they need to film the entire episode so that other scientists can learn something. So, if you ever meet me and I'm complaining of a stomach problem, cameras at the ready please.

CHAPTER TWELVE

Nine Elements that Changed the World (and One that Didn't)

THE LONGEST EXPERIMENT IN HISTORY

Classifying something as a solid, a liquid, or a gas is usually straightforward. Solids don't flow, liquids do but can't be compressed, and gases are both compressible and capable of flowing. These definitions work for most materials but there are some that aren't what they first appear, bringing us to our final record-breaking chemical: pitch.

Also called asphalt, pitch is the sticky black residue left over when crude oil is distilled. We use it to make our roads and what makes it interesting is that while it appears solid, it isn't. The roads you drive on are made of liquid.

In 1902, an unnamed scientist at the Royal Scottish Museum in Edinburgh poured a sample of hot pitch into a glass funnel and left it to cool. For over a hundred years the pitch has oozed through the funnel and two drops have fallen onto a dish below.[1] To the naked eye it looks like solid black gunk, but what you're looking at is the most viscous liquid known to humankind.

A similar version with a slightly runnier pitch was set up in 1927 at the University of Queensland in Brisbane. That one has dripped nine times since the experiment began, with the most recent one falling in 2014.

Time-lapse cameras have captured the slow creeping of these liquids, but nobody has ever witnessed the precise instant when a drop falls. Don't despair, though. If you go to http://www.thetenthwatch.com/feed you can watch a live broadcast of the Brisbane experiment as the tenth blob of liquid slowly forms. You're welcome.

These two experiments have been running through both world wars, the rise and fall of the Soviet Union, and the release of every single *Fast and Furious* movie, making them the longest running experiments in history. But if we wanted to get philosophical for a moment, we could argue that one experiment has been running for even longer and we are right in the middle of it.

What happens if you take a planet's worth of elements, clump them into a ball orbiting a backwater star, and leave the whole thing for 4.5 billion years? What will happen within the planet's core and what will happen on its surface?

Humans are a latecomer in a long line of chemical reactions carried out with elements that have been around since before the dinosaurs roamed. The story of the elements is also the story of us and the periodic table has been there for every step, whether we knew it or not.

So, in the final chapter, I want to examine which of the elements have been crucial to our development and which ones have had the biggest impact on this experiment called humanity.

COME BACK ZINC!

There's an episode of *The Simpsons* where Bart is forced to watch a video about a kid called Jimmy who wishes to live in a world without zinc. He soon discovers his car battery no longer exists, preventing him from picking up his girlfriend Betty. Not only that, the rotary mechanism on his phone has vanished, as has the firing pin in the gun with which he tries to commit suicide. Jimmy suddenly wakes up screaming, "Come back zinc!" and breathes a sigh of relief. It was all a terrifying dream.[2]

It's a perfect satire of the hokey educational videos popular in the 1950s, because nobody has ever wished to live in a world without zinc. I do know someone who considers zinc her favorite element, but most people probably know little about it.

And that's true for most of the elements on the table. We know they exist but don't give much thought to what they do. If you suffer from kidney problems then you should thank zirconium because it's used in dialysis machines for absorbing ions. If you're a smoker you owe your habit to cerium because it's one of the only metals that produces sparks, allowing your lighter to function.

If you work in welding, your goggles are tinted with praseodymium to block yellow light. Or perhaps you work in the solar-panel industry; if so, ruthenium is the element to get excited about because it absorbs sunlight better than anything else.

The microwave you use to heat your meals wouldn't function without samarium. The fountain pen you used in school had a nib made of iridium and if you live in mainland Europe the dollar bills you spend are impregnated with europium to detect forgery.

Everyone will have their own favorite element (and if it's not phosphorus, what's wrong with you?) but we can make a case for some having played a more important role than others.

We could argue that aluminum has been more important than selenium, for instance. One is used in construction and vehicle manufacture while the other is used to decolorize glass and eliminate dandruff. (Having said that, I do enjoy looking through my windows while running my fingers through a fine crop of healthy hair.)

If we ignore obvious and boring choices like the oxygen we breathe or the iron in our planetary core, which elements have played the most crucial roles in our cultural, political, and technological evolution? Which ones have made the world what it is, and which ones are secretly influencing our daily lives without us even noticing?

This has been a difficult list to compose because as soon as I settled on one selection I immediately felt I was leaving an important element out. The problem is that every element is special. Well, all except for one.

AN HONORARY MENTION

Originally I intended this chapter to be a conventional top-ten list, but in the end I went for nine. The reason is that there's one element that deserves a very special mention but doesn't quite fit with the others.

In the process of researching this book I learned the stories and characteristics of all 118 known elements. Every one is unique either because it played an important role in the history of chemistry or because it has a distinct property making it ideal for a particular use.

I have succeeded in namechecking every element somewhere in the book at least once, with the exception of element number 66—dysprosium. The most pointless element in the world.

Dysprosium was isolated by Paul-Émile Lecoq on his mantelpiece in 1886 and that seems appropriate.[3] It's a mantelpiece element if ever there was one. It exists and it probably has a purpose, but nobody knows what it is.

Dysprosium is neither especially rare nor especially common. It reacts with water but not as well as group 1 metals. It can be used to make lasers but they're not as good as those made from helium or neon. It's occasionally used in nuclear control rods, which stop things getting too hot, but you can achieve the same effect with indium or cadmium. Dysprosium is beaten at every turn by something else.

There will definitely be a dysprosium scientist out there who's currently foaming at the mouth as she reads this. But dysprosium doesn't seem to be exclusive in any way, which makes it quite interesting.

I hereby declare dysprosium to be the only element you could remove from human history and absolutely nothing much would change. We salute you, dysprosium, the most boring element on the periodic table.

Right, on with the list.

ELEMENT OF AGES

Carbon is an obvious choice to kick things off. It's so vital to our world it's practically humdrum. Look around the room and probably 90 percent of things you're looking at are either made from, extracted with, or powered by carbon. It is the element that has defined the ages of humankind.

We've been around for hundreds of thousands of years, but what we call civilization began with manipulating metals. The Stone Age represented the primitive infancy of our species but it was the Bronze and Iron Ages that were the turning points.

Before we mastered the art of metallurgy, the only metals we knew about were gold and occasionally silver, so all our construction materials, weapons, and tools came from bashing rocks together. Then at some point between 8000 BCE and 3000 BCE everything changed.

Most metals in nature are bonded to oxygen, but oxygen forms better bonds with carbon. This means if we mix enough carbon together with our metal oxide (rock) and give the whole thing some energy (heat it), everything rearranges to carbon dioxide and pure metal. This technique, known as smelting, was the most important chemical reaction since fire itself.

The early technologists, whoever they were, discovered that roasting rocks in the presence of charcoal produced metal. First, we began extracting copper and tin, giving us bronze. Then we learned how to get the fires hotter and started extracting iron, previously only found in meteorites.

By the nineteenth century we were burning carbon itself as a fuel source, using it to run our combustion engines. Carbon has an advantage over other fuels because, rather than leaving unpleasant residues, it burns away to an invisible gas. Where's the harm in that?

Today, we still use coal for our power stations, so the electricity you use is most likely down to carbon too. It's only in the past sixty years we've realized that all that CO_2 has the awkward feature of absorbing infrared radiation, slowly heating the atmosphere as the decades tick by.

On the plus side, carbon is also the basis of polymer chemistry. Take a long chain of carbon atoms, use hydrogen to make sure each one has the correct number of bonds, and if you tangle the ropey chains together you end up with a plastic.

Imagine a world without plastic, metal, or widespread electricity and you begin to see why carbon is so important.

Carbon's versatility is a result of its location on the periodic table. It sits on the top row, making it a small atom capable of forming tight bonds, in the fourth column along, giving it four available bonding electrons.

An element like fluorine is also on the top row but it is only one electron away from a filled shell, meaning it will form one bond and then stop. Carbon has four electron spaces, meaning it can form four links to other atoms, all of them strong.

Other elements that form multiple bonds are usually too big for the bonds to be robust, so carbon has the best of both worlds, which is why we find it in everything from our cell membranes to our cell phones.

It gave us the materials we use and the power to manipulate them, and now its presence in the air is threatening to knock our climate out of equilibrium. If there is one element that has turned the course of human history more than any other, it's carbon.

FOOD FOR AN EMPIRE

At the start of the 1800s Britain's armies were expanding across the globe. The Napoleonic wars were finishing, slavery was coming to an end, and the Empire was approaching its "golden age." But the admirals and generals of this ruthless military machine were facing a problem. An empire is only as strong as its food supply. How do you get food to thousands of people, far away from where it's being produced?

The answer was discovered by a French inventor named Philippe de Girard who devised a method to vacuum-seal food in a tin can. After testing his invention on several British scientists, the idea was sold to the engineer Bryan Donkin who set about improving the method.

Donkin was already a superb craftsman who consulted on the manufacture of Babbage's difference engine and Telford's suspension bridge, and was also the inventor of the humble pen. While the inventor of the pen is sometimes misattributed to John Loud in 1888, Donkin already had a patent in 1803.[4] Let's just get our pen history right, folks.

By 1813, Donkin had designed a method to mold tin cans in such a way that food inside would be locked in without any air, meaning it could last for years and be transported as far as was needed.

After Queen Charlotte sampled a tin of his corned beef and praised the taste, Donkin began manufacturing tin cans en masse and sold them to the Navy. Tin cans allowed countries to feed their armies during both world wars and today over forty billion are sold around the world annually.[5] While many of these cans are made from steel today, it's the tin plating that prevents irreversible rust.

What also makes tin special is not what it does as a pure metal, but how it can modify other metals when they are mixed together, forming what's called an "alloy."

Its softness is one of the reasons it is mixed with copper to make bronze. When alloyed with lead it forms pewter, the material most cutlery was made of until very recently. When mixed with a bit more lead you end up with solder, the "glue" used in electronics to join wires.

Bell metal, used for making bells, is an alloy of tin with copper. Gunmetal used for making, well, guns, is an alloy of tin with copper and zinc. Terne, used to make roofing, is tin mixed with lead again. Ball bearings are usually made from tin with copper and iron. Galinstan, used for telescopes, is tin mixed with gallium and indium, and the list goes on. Tin is the great modifier of the periodic table.

It's not quite as prevalent as iron but it has the clear advantage of being rustproof and, because it's easily extracted and manipulated, anyone can work with it from the richest monarch to the lowliest commoner. While armies and politicians might have prized elements like gold, tin has always been the element of the people. Not that gold hasn't been important, too, mind you.

ALL THAT GLITTERS

The color of gold has led many cultures throughout history to worship it, often associating it with the sun (silver being linked to the moon). It arises because gold has large gaps between the atomic orbitals, so visible light loses a lot of energy when it strikes. The highest energy colors like violet, blue, and green are absorbed into the metallic surface while the yellows and oranges are bounced back out. Cesium and copper also have yellow/orange hues, but nothing compares to gold.

As we saw in Chapter 3, gold was essential to discovering the nucleus and therefore modern chemistry itself. It was used because it's the most malleable metal available, so soft that 28 g would be enough to make a wire stretching nine times the height of Everest.[6]

This ease in molding, as well as its shine, has also led to its use in jewelry since prehistory, not to mention the fact it doesn't tarnish. While other metals will gradually react with oxygen, gold will gleam forever.

It's also a very rare metal. If you were to collect all the gold deposits in the world it would total around 170,000 tons. That would barely fill three Olympic-size swimming pools.[7]

This combination of malleability, rarity, permanence, and beauty are what make it so precious. Gold can be traded anywhere in the world, regardless of local custom, because everyone values it.

In Finland the skins of squirrels used to be acceptable as money, and up until the twentieth century Ethiopia used blocks of salt.[8] Money is different wherever you go, but gold is revered everywhere and always has been, making it the only true international currency.

Alexander the Great led the Greek army to conquer the Persian Empire—the largest in the world—in order to steal their gold. Julius Caesar did the same thing to western Europe. So did King Ferdinand of Spain, sending his conquistadors to rip gold from the Americas (and we all know how that story turned out).

The first gold coins were used in China during the sixth century BCE, but by the 1800s every large country in the world (apart from China, ironically) was using a gold standard for international as well as domestic business.

Owing to its rarity and weight, though, gold coins are far from practical so banks began printing contracts that corresponded to a certain amount of solid gold. This was the invention of modern money itself.

KNOWLEDGE AND POWER

Some of the elements have a split personality. The same substance can be of great benefit to the world, but also the cause of endless pain. No other element can lay claim to having enlightened so many or killed so many as lead.

Once extracted from its ore, lead is a dull metal with three important properties: density, meaning it's hard to break, malleability, meaning it can be bent, and corrosion resistance, meaning you can have it in contact with water.

The Romans carried out lead mining on a grand scale because they used it for pipes and waterworks. Iron is no good because it rusts so lead was used at a rate of thousands of tons per year. The very notion of water straight to people's homes hadn't been explored properly until the Roman plumbing system. Even the word plumber comes from the

Latin word for lead, *plumbum*, because piping specialists were plumbum experts. That's why it has a silent "b."

Because of its toxicity, some people have speculated that lead-poisoning contributed to the decline and ultimate defeat of the Roman Empire.[9] This seems unlikely, however, as lead poisoning was already a known malady and water doesn't usually dissolve enough to reach dangerous levels.[10]

It's possible that boiling grape juice in huge lead vats may have caused lead poisoning in some of the aristocracy, but this is speculation at best. It's unlikely that lead caused the collapse of Roman civilization, but don't worry, it's still responsible for millions of deaths every year.

In thirteenth-century China, it was realized that a small tube of gunpowder could launch a projectile at high velocity when it exploded—the invention of the gun. The technology spread to European armies and the best metal for making bullets turned out to be lead—not only because it's readily available and easy to manipulate, but because it is so dense that once it is fired from the barrel it keeps going in a straight line. No other metal allows us to shape it so well while being dense enough to hold its trajectory.

Nobody knows how many bullets are manufactured in the world today but the number is probably in excess of ten billion a year: enough for one bullet per person. It's hard to think of a weapon that has caused more death than guns firing lead bullets.

But lead has also done wonders for us. In 1440 Johannes Gutenberg was looking to find a way of quickly conveying information to people. Up until then, every text and book had to be copied by hand. If a machine could be rigged to do the job, books could be produced in a matter of days rather than months.

The result was his printing press, only achievable thanks to lead (alloyed with a little tin). Because lead was so malleable, it could be carved into the precise shapes of block letters. Other metals could be molded too, but lead's density meant hammering it repeatedly onto a page wouldn't cause it to wear away.[11] The same characteristics that help lead kill are those that help it educate.

DRINK IT DOWN

People are living longer these days. Obviously a good thing. The only downside is that we're more prone to age-related disease. This has led to a lot of hoo-ha and fear-mongering about the apparent rise of cancer and heart disease. I've heard everything blamed from GMO foods to (perversely) chemotherapy drugs themselves, but it really comes down to cold numbers.

Humans die. Sorry to break that to you. Our bodies are fragile and they aren't built to last. The older you get, the less you tend to function and the more likely you are to die from something like cancer or heart disease. The only reason we've seen an apparent rise in these deaths is because people are lasting long enough to die from them. Age-related illness has existed as long as the human body; it's just that most people tended to extinguish before they got that far.

Death is always unpleasant but I would say age-related illness is a fair price to pay for a life expectancy in the eighties. During the mid-1800s, life expectancy was forty-two, mainly because people died in childhood, bringing the average down.[12] The only reason we enjoy a higher number today is simple. It has little to do with a gluten-free diet or a Pilates class. It's because we have defeated the world's number-one killers. We don't die of infection anymore.

In the 1340s, hundreds of millions of people died from bubonic plague. Between 1817 and 1917, an estimated thirty-eight million died from cholera.[13] Measles and smallpox have been responsible for more deaths worldwide than any war you care to mention and don't get me started on polio or malaria.[14] In many parts of the globe, these diseases are still rampant, but in the West we are fortunate because we have eradicated them. Quite frankly, dying from old age is something for which we should be grateful. Many are not so lucky.

The reason we aren't seeing epidemics breaking out every year is down to two things: vaccination and element number 17, chlorine.

The first widespread use of chlorine was during the First World War when the German chemist Fritz Haber introduced it as a chemical weapon. In 1915, he oversaw the installation of five thousand canisters of it along a 7-km distance at the western front and, when the wind began blowing the right way, Haber ordered the canisters be opened.

Chlorine is a thick green gas that rolls along the ground like a liquid. Carried by the wind, the chlorine was dragged toward the British Army and flooded their trenches, asphyxiating and blinding thousands of men.

According to Hermann Lutke, on May 1, 1915, a party was being held in Haber's honor to praise his simple but effective use of chlorine chemistry. A few hours after the party, his wife Clara (a noted pacifist) took Haber's service revolver into the garden and shot herself in the chest, dying moments later in her son's arms.[15] In this context, chlorine has a similar reputation to lead but, just like lead, it can be put to a far better use.

Because it is lethal to biological organisms, if handled correctly it can be used to kill the pathogens that would otherwise be lurking in our water supplies.

The average person in the US and the UK uses around 340 liters of water a day and it has to be clean in order to stop the spread of disease.[16] Even toilet water has to be potable because if it contained anything harmful it could get airborne during flushes.

There are a few alternatives to chlorination, such as bubbling ozone through the water, but chlorine is the main choice for every European country and the whole of the United States.

It works because chlorine dissolves to form hypochlorous acid, $HOCl$, which is lethal. That's what kills you if you are unfortunate enough to inhale it as a gas. It's easy to remove from water, though, so if we pump it into our drinking supply it will kill everything, after which we remove the excess with carbon charcoal.

While adding fluoride to the water has caused controversy (largely because it was implemented before long-term studies were finished), nobody objects to chlorine. It's the main reason you aren't currently dead.

THE SILVER SCREEN

Any writer of non-fiction, no matter how objective they claim to be, will be writing with bias, putting their own personal views into things, often without realizing it. And by the way, don't you just hate celery? Almost all of recorded history has been the result of eyewitness accounts and people's memories, which makes things hard to verify.

That changed in 1717 when a German chemist named Johann Schulze left a bottle of silver nitrate and chalk on his windowsill. Schulze placed the bottle down absentmindedly and, when he picked it up a few minutes later, was shocked to find it had turned brown. Except for a thin white line suspended in the liquid.[17]

He looked out the window to see what could have reacted with his solution and noticed a piece of thread hanging across the window in exactly the same shape as the white line inside the bottle.

Where the sunlight hit the silver nitrate it made it go dark, but where something had obscured the sun, the liquid remained white. Schulze had taken the very first photograph and it was a liquid. Considering Henri Becquerel's radioactivity discovery, it's strange how many times leaving a jar lying around has led to a monumental realization.

Silver atoms can be bonded to nitrate molecules in solution, but given a bit of energy they can separate and form solid metal. A lump of solid silver glistens brilliantly but powdered silver is a dark brown, showing exactly where the light has hit it.

It was a French inventor named Joseph "Nicéphore" Niépce who realized that putting silver compounds onto a piece of paper and focusing images with a pinhole camera would create a black-and-white copy of what was being projected. In 1829 he used this technique to capture the world's first proper photograph from his bedroom window, *View from the Window at Le Gras*, taking eight hours of exposure time.

At least, that's the official story. In 1777, another scientist had already discovered that you could trap silver-solution images on a piece of card using ammonia. This scientist also figured out the cause of the phenomenon but never pursued the research, denying himself the title of being the inventor of photography. That scientist (and I'm not making this up) was none other than Carl Scheele.

Over the following century we discovered other silver chemicals that reacted faster than nitrate and, by using lenses, we were able to intensify the light, creating instantaneous images of particular moments. We didn't have to rely on word-of-mouth or written accounts to store information anymore: silver allowed us to capture pictures of things as they truly were.

Historians disagree over who took the idea of photographs and strung them into movie reels, but the patent for the first film camera seems to have been filed by one Wordsworth Donisthorpe in 1876.[18] He used it to film a few seconds of Trafalgar Square and started the movie industry, which we still call "the silver screen" after the element involved.

Color photography relies on silver as well but with additional chemicals that respond to different frequencies of light. When red light hits the layer of film containing the red-sensitive chemical, it causes silver powder to form. The same happens in the blue layer with blue light, and the green layer with green light.

You end up with a black-and-white image of where the reds should be, one for the blues, and one for the greens. By developing each layer with the correct dye in the right order, you can recreate the color image with which you started.

It's hard to imagine a world where photographs or film didn't exist because they are so ubiquitous. We rely on them to give us reliable information and silver is what made it possible. In recent years the invention of computer editing technology and digital cameras has changed things somewhat, but there was originally truth to the phrase "the camera doesn't lie."

DESTROYER OF WORLDS

For a long time, the science of nuclear bombs was highly classified. When Nobel Prize–winning chemist Linus Pauling gave a public talk on the subject, an FBI agent showed up at his office to interrogate him on how he knew the workings of a bomb so perfectly. Pauling responded, rather coolly, by saying, "Nobody told me, I figured it out."[19] Nowadays, the design of an atom bomb is well known and it's all about uranium.

Uranium atoms have 92 protons in their nucleus and usually 146 neutrons, giving a total of 238 particles. But roughly 0.7 percent of uranium atoms have 143 neutrons instead, making uranium-235. This combination of protons and neutrons is volatile, and when the nucleus fractures it spits out neutrons. These neutrons go flying off and get absorbed by other uranium nuclei, making them unstable and causing another fission event.

If you've got 1 kg of uranium, most of your neutrons can escape through the surface of the metal, but once you get to around 47 kg, what's called critical mass, the neutrons in the center don't escape. The energy builds, fissions multiply, and the result is a nuclear blast.

It might be fair to say that gold dominated global politics up until 1945, but uranium certainly dominated it afterward. With 47 kg, you can end one war and start another.

On August 6, 1945, a uranium bomb was detonated over Hiroshima, causing the deaths of over eighty thousand people. Three days later, a plutonium bomb (made from uranium as a starting agent) was dropped on Nagasaki, killing forty thousand people and bringing the Second World War to a close.

With the ability to manipulate uranium, the United States became the most powerful nation on Earth. That kind of strength invites challenge. Four years after Hiroshima and Nagasaki, the USSR demonstrated its own nuclear capability and the Cold War began, shaping the technological, cultural, and economic landscape of the twentieth century.

Most nuclear weapons today are based on plutonium but uranium is still the starting ingredient. Getting hold of it isn't difficult, mind you. Uranium was used for Fiesta dinnerware glazes (hilariously the US government confiscated the products during the Cold War). The tricky bit is extracting the 0.7 percent of atoms that are fissile.

At the time of writing, nine nations have the technology to do this, with the United States and Russia owning the largest stockpiles. The precise number of warheads is unknown but it's estimated to be well over five thousand each.[20] With that arsenal, you could eliminate all life on Earth many hundreds of times over.

The physicist who coordinated the invention of nuclear weapons was Robert Oppenheimer. He was once asked in an interview what it was like to witness the first nuclear-bomb test, codenamed Trinity. His response was chilling:

We knew the world would not be the same. A few people laughed, a few people cried. Most people were silent. I remembered the line from the Hindu scripture the *Bhagavad Gita*. Vishnu is trying to persuade the prince that he should do his duty and, to impress him, takes on his multi-armed form and says "Now I am become Death, the destroyer of worlds." I suppose we all thought that. One way or another.[21]

WE WANT INFORMATION . . . INFORMATION . . . INFORMATION

Silicon sits right below carbon on the periodic table and has a similar electronic structure. The only difference is that it's bigger, so its bonds are not as strong.

It can form crystals similar to diamond with comparable strength, and can also be strung into plastic chains, the most famous of which is the silicone gel responsible for some people's career success in Hollywood. But silicon's primary use is at the nerve center of every electrical device you own.

If the nineteenth century is remembered for the Industrial Revolution and the internal combustion engine, the twentieth century should be remembered for the silicon revolution and the transistor, an invention of which many have never heard.

Invented in 1947 by Walter Brattain, William Shockley, and John Bardeen (the only person to win two Nobel Prizes for physics), transistors are to computers what bricks are to houses. Your smartphone contains about three billion of them and your laptop contains seventy times that.

A transistor's job is to let electrical current pass through it sometimes and block it at other times. On its own, this sounds mundane, but get enough transistors hooked up in an intricate pattern and you've got a microchip. By programming a series of instructions for these transistors as 1s and 0s, we can tell transistors to switch currents on and off, allowing us to control circuits and store information.

The problem with making a transistor out of metal is that metals always conduct. Similarly, non-metals are always insulators. In order to create something capable of switching on and off at different times you need an element that is halfway between a metal and a non-metal. Enter silicon.

Silicon atoms are large so they're vaguely metallic in nature, but their shape has more in common with non-metals like carbon and boron. These hybrid properties make silicon a semi-conductor and its crystals form the backbone of transistors.

Not only that, silicon is also the key ingredient in glass, giving us the optical fibers for the internet. Not to mention making windows.

Most optical fibers are made by one company called 3M and their glass is so transparent that, if you were to make the ocean out of it instead of saltwater, you would be able to see to the bottom with perfect clarity.

During the 1950s, after inventing the transistor, William Shockley set up a business in California doing research with the computer science department of Stanford University.[22]

Prior to his invention, all computers were mechanically based and occupied whole rooms. Silicon offered the possibility of computers you could have on your desktop.

Once interest in silicon began to boom so did the local economy, and today Shockley's neighborhood is the headquarters of Apple, eBay, Facebook, Google, Intel, Netflix, Yahoo, and Visa. It's a region of southern San Francisco called the Santa Clara Valley, more commonly known by a name inspired by the element that built it: Silicon Valley.

Silicon enables us to perform calculations that previously took a library of people days to complete and runs everything from our digital watches to our mobile phones, although that technology comes with a moral dilemma tied to a different element—tantalum.

Tantalum vibrates when electrified, making its importance in mobile phones obvious. Seventy percent of the world's tantalum deposits come from the Democratic Republic of Congo, a country whose economy is based on its mining and export. The civil war that raged there from 1994 to 2002, the bloodiest conflict since the Second World War, was funded through the sale of tantalum.[23] Sometimes our relationship with the elements is ethically quite dark.

SAVIOR OF WORLDS

In the 1930s, hydrogen was set to be the element of the future. It's easy to get ahold of, easy to transport, and when it burns the only by-product is water. It's the cleanest, greenest fuel imaginable.

Not only that, its low density makes it perfect for generating lift. An airplane needs to build a lot of speed for its wings to bite the air, but a hydrogen dirigible will float without assistance or persuasion. Helium is less reactive, which makes it safer, but when the United States began stockpiling it in 1925 at the National Helium Reserve in Amarillo, European agencies turned to hydrogen as the obvious alternative.

The German government was particularly eager to harness hydrogen technology and in 1931 began constructing the world's largest zeppelin, LZ-129 *Hindenburg*, a marvel of chemical and aeronautical engineering.

But on May 6, 1937, as it was being tied to the ground of Lakehurst Naval Air Station, the *Hindenburg* caught fire. Nobody knows how it started (spontaneous zeppelin combustion?) but in under thirty seconds all two hundred thousand cubic meters of hydrogen had combusted.[24]

The crash was caught on camera and the accompanying audio by Herbert Morrison shouting "Oh the humanity!" has become iconic. People saw what a hydrogen fire looked like and the age of the zeppelin was over before it began.

The world didn't hear much from hydrogen for a few decades, until the USSR detonated the Tsar Bomba in 1961 (see Chapter 4). The Tsar wasn't any old uranium bomb: it was a hydrogen bomb and the difference was obvious. With a mushroom cloud reaching 64 km into the sky, it made the bombs dropped at the end of the Second World War look like fire crackers.

The exact details of how a hydrogen bomb works are still classified and, bearing in mind what happened to Linus Pauling, I'm reluctant to do extensive research on it. While writing this book, I've investigated the price of plutonium and how much thallium is needed to kill someone. I should probably exercise caution before I start asking people how to build an H-bomb.

The basic premise, though, is fairly well understood. Einstein's $E = mc^2$ equation tells us that we can obtain energy from an atom by splitting it. What's surprising is that reversing the process and fusing nuclei together releases even more energy (because quantum mechanics, that's why).

The bomb works in two stages (I think). First, a conventional uranium bomb is triggered and the heat from that blast causes a capsule of hydrogen atoms to fuse, generating a miniature sun as they convert to helium. That's the awesome power being demonstrated in images of the Tsar Bomba explosion.

Combined with the terrifying sight of the *Hindenburg*, hydrogen has been an element of terror in the public eye. But we shouldn't give up on it. In fact, as the future creeps toward us, we may find ourselves becoming entirely reliant on it.

The energy released from fusing hydrogen doesn't necessarily have to be done in one go. In the same way uranium rods can be brought near each other to generate heat rather than explosions, it should be possible to bring hydrogen nuclei together in controlled conditions.

Fusion-based nuclear plants would produce no toxic products, would end our dependency on fossil fuels, would end all conflicts fought *over* fossil fuels while providing limitless energy for the planet, as well as bringing an end to human-made climate change. Fusing hydrogen could really be the ticket humanity needs to solve all of its problems. There's only one slight snag—we haven't been able to do it.

In order to fuse hydrogen atoms, you have to heat them fast enough to collide. That takes energy, and all our current fusion reactors take more power to get going than we can usefully extract.

We have only achieved one positive fusion reaction so far, at the National Ignition Facility in California in 2013. There, a group of researchers led by a man whose name is genuinely Omar Hurricane, blasted samples of hydrogen with laser beams and excited them into fusing. Mr. Hurricane and his team are the first and, to date, only people to have successfully got more out of a fusion reaction than they put in.[25] It's not perfect and it's not enough to power the world, but it's a promising step.

And there's something that may be even more important. Because hydrogen burns beautifully with oxygen, it's a perfect rocket fuel. Those enormous tanks you see on the sides of ships being launched into space aren't full of petrol, they're full of chemicals that generate hydrogen and oxygen.

Hydrogen isn't just the element which may save the world, it's the element which may help us leave the world altogether. And sooner or later, we're going to have to.

At the moment we're living in a golden age of pulling elements out of the ground with abandon, but it can't last. Assuming our planet doesn't get obliterated by an asteroid (we're overdue), we will eventually consume all the resources Earth has kindly given us.

If our species wants to survive, we're going to have to do it somewhere else, which means we need to get out there and exploring. For that, we're going to need hydrogen, our ticket to ride the universal express.

A FINAL THOUGHT

Each element on the periodic table has a story to tell but what that story is depends on us. It is our duty not to abuse such power. And I don't think we will.

When I look at the periodic table, I see a monument to how far we have come and how much we have learned in such a short time. Through science we are capable of understanding the Universe and using its resources to do amazing things. I truly believe that science will save our species.

Sulfur with an "f"

Naming an element is usually an honor given to the person who isolates it. Unfortunately, it can cause disagreements when scientists give elements unpopular names.

In 1875 the French chemist Paul-Émile Lecoq named a new element gallium from the Latin *gallia*, meaning France. However, it was soon suspected he had been a bit sneaky. *Gallus* is also the Latin for the rooster, which translates into Lecoq, his own name. Perhaps he had immortalized the Lecoq name by subtly naming the element after himself.

To try and solve these problems, the International Union of Pure and Applied Chemistry, IUPAC, has stringent rules for the naming of a new element. Elements can be named after:

1. a character from mythology (e.g. thorium, after the Norse god Thor);
2. a place (e.g. rhenium, from *Rhenus*, the Latin name for the Rhine river);
3. a property of the element (e.g. bromine, from the Greek *bromos*, meaning foul stench);
4. the mineral from which it was extracted (e.g. samarium, after the mineral samarskite);
5. a scientist (e.g. roentgenium, after Wilhelm Röntgen the discoverer of X-rays).

IUPAC will deliberate a proposed name for up to five months before giving the thumbs up and then, once they have spoken, the name is internationally recognized and periodic tables are adjusted to incorporate it.

Many British chemists were horrified in 1990 when IUPAC endorsed the American spelling of sulfur with an "f" as opposed to the British spelling of sulphur with a "ph." And throughout this book, I have gone along with their decision.

To be clear, IUPAC is well within its right to favor sulfur over sulphur. The etymology of the word is unknown and anyone who claims otherwise is misinformed. The first recorded usage is to be found in the writing of the second-century BCE poet Ennius, who called it *sulpureus*. That word itself (whose etymology has been lost), however, may come from the word *swefel* (whose etymology has also been lost).

As we don't know where the word comes from, there is no reason to prefer one spelling over another. Spelling it with a "ph" isn't just a matter of British pride: it's refusing to accept the agreed international standard.

Personally, I find it infuriating I have to write the name of an element with a lower case and my own name with a capital, but I have to play ball (in this book, at least; on my website I capitalize whatever I feel like!).

I have heard some people suggest IUPAC accept a trade where sulfur is spelled with an "f" in exchange for aluminum being spelled with two "i"s (the British way, aluminium). It's maybe worth pointing out that Humphry Davy, the English scientist who named it, did originally choose aluminum so the American spelling is more authentic after all.

Half a Proton?

The closer we look at particles the more substructure we find. Atoms are made of electrons and a nucleus. The nucleus is split into protons and neutrons. How do we know when we've really got to the bottom of it?

During the 1960s, theoretical physicists decided it was time to take a bottom-up approach rather than a top-down one. Starting with the basic laws of nature, what fundamental particles should we see arising? The resulting framework, called quantum field theory, predicts a buffet of particles, all of which have been found, so the approach is definitely along the right track.

Electrons turn out to be fundamental, as do photons, the particles that make up light. There are lots of others with names like neutrinos, gluons, and Higgs bosons, but protons and neutrons are not on the list.

It turns out that protons and neutrons themselves are not fundamental but can be thought of as three particles that are. A proton can't be split in half but it can be described as being made up of thirds. Murray Gell-Mann named these particles quarks (pronounced "kworks" not "kwarks").

It still wouldn't be right to say you can chop a proton into thirds, however, because they don't actually let you do that. Quarks do not exist as individual things, but in little pairs and trios.

If you were to take a proton, made from three quarks, and break it apart you wouldn't end up with the three individual quarks, you'd end up with six . . . that's quantum field theory, folks.

So, while in one sense you can describe a third of a proton, you could never actually have it. The quarks never leave their proton so it's fine to talk about protons as if they were fundamental particles. They might as well be!

Schrödinger's Equation

Schrödinger's equation is a full description of everything we can know about what a particle is doing. We might be interested in looking at how a particle is going to behave at a specific point in time or at a specific point in space. We might not be interested in either and only want to know what energies are involved or what rotations a particle can have.

This means there are lots of different forms of the Schrödinger equation and different ways of writing it. The most straightforward one is called the generalized time-dependent Schrödinger equation and it looks like this:

$$H|\Psi\rangle = i\hbar \frac{\partial |\Psi\rangle}{\partial t}$$

i represents the square root of minus one. Any normal number, either positive or negative, always generates an answer that is positive when multiplied by itself. -2×-2 isn't -4, it's $+4$. But that means -1 would have no square root, so there must be another type of number which multiplies by itself to generate negatives. These numbers are called i numbers. The reason the letter i is chosen will muddle things at this point so don't worry about it: it's a historical convention and it's meaningless. It may seem strange that we have to use freaky numbers in our equation because it seems like a cheat but that's not what's happening. Nature does things outside of our normal experience so we have to use numbers outside our normal experience in order to make sense of it. When we try using regular numbers, the equation gives answers that don't match reality. It would seem that nature uses i numbers so we have to as well.

H is called the Hamiltonian and it refers to the total energy of the thing we're examining. We've written it here with one letter but that's a shorthand. Written in full, the Hamiltonian is a long-winded term that takes into account the particle's mass, its kinetic energy, how far it is from the nucleus (called its potential), and so on, but it still just means how much energy the particle has.

Ψ: This symbol (psi) represents something called the wavefunction. It can mean a great many things, but in the context of chemistry it refers to the fact that the probable location of a particle ripple. Rather than having a specific coordinate in space, an electron's location has a wave-like character when it's not being interfered with. The wavefunction takes this into account.

$|>$ is called a ket vector. What it refers to, generally speaking, is the state that something is in. In this case the state of the particle's wavefunction. The left side of the equation, read in full, is now telling us that if we calculate the total energy of the wavefunction's state and multiply the answer by negative i, we'll get something useful.

$\hbar$ is called Planck's constant and represents 1.055×10^{-34} Joule-seconds. This number is a property of the Universe that relates the energy of a particle to its frequency. Frequency is how many times something will ripple per second and since all particles have ripply movement, we need a term that relates the two. Specifically, if we divide the energy of a particle by its frequency, we get a number called h, which is 6.626×10^{-34} Js in SI units. Planck's constant is arrived at by dividing h by 2π. We do this because 2π is often used when measuring frequencies, so we include it as part of our constant to make the equations neater.

∂ is called a partial differential. It's a symbol that tells us to measure how one property changes when you've got lots of other stuff going on and you only want to focus on one thing. In this case $\dfrac{\partial}{\partial t}$ is telling us to compare the change of something compared to t (time).

So, the whole equation is telling us that if we can work out the total energy a particle has in a particular state (left-hand side), we can work out how its behavior will change with time (right-hand side).

If you know what energy an electron has, you can predict where it's likely to be at any moment. Do this for all three spatial dimensions and you'll end up with a description of where an electron is likely to be around its nucleus—the orbital.

Neutrons into Protons

We've already met quarks in Appendix II and they can be tricky things. They come in many varieties but they all have an electric charge that adds up to the charge of a proton or a neutron.

An "up" quark has a charge of +2/3 while a "down" quark has a charge of –1/3. When two up quarks and a down quark occur in a trio, the charges combine to create an overall +1, a proton. If an up quark occurs with two down quarks, however, the charges cancel and the result is a neutron.

But quarks don't stay as one type. An up quark can turn into a down quark and vice versa. A neutron is an udd (up down down) combination, but if one of the downs turns into an up we get an uud (up up down)—a proton. It's the quark inside that flips and turns a neutron into a proton.

When this happens the amount of overall charge has changed and for some reason this is a big no-no for the Universe. Rather than creating a charge imbalance, the Universe prefers to keep things neutral so it does a particle shuffle.

When the –1/3 down quark changes character, it emits a particle called a W⁻ boson, which carries away a –1 charge, leaving a +2/3 charge behind.

The W⁻ quickly splits into an electron, which retains the charge, and another particle called an anti-neutrino, which has none. And there's no simpler way of describing the whole process.

APPENDIX V

The pH and pK$_a$ Scales

You may have met the pH scale in school. The more acidic something is, the lower its number. Acids tend to have values below 7 while non-acidic things tend to be 8 and upward. The reason the scale was introduced was because the numbers involved in acid chemistry are often extremely small.

Suppose we had 1×10^5 hydrogens in a 1-liter bottle and 1×10^4 in another. The first contains 100,000 and the second contains 10,000. Clearly, the first is ten times more concentrated than the latter, but they are both extreme numbers. So we invoke the laws of logarithms.

A logarithm is the number of times you have to multiply something by itself in order to get a particular result. Say you had the number three and multiplied it by itself four times. That would be written as $3 \times 3 \times 3 \times 3$ or, more simply, 3^4. The answer is 81.

But suppose you want to do your calculation the other way around. You want to know how many times you had to multiply three by itself to reach 81. You would write it like this:

$$\text{Log}_3 81 = 4$$

In other words, what number do I have to raise three to in order to reach 81. The answer would be 4.

In our earlier example, we had one solution containing 100,000 hydrogens. If we express this logarithmically we would write:

$$\text{Log}_{10} 100,000 = 5$$

Five is a much easier number to work with so we might describe this acid as a "5 solution." The number isn't telling us the concentration directly but it's telling us the order of magnitude with which we're dealing.

Likewise, a solution ten times more dilute would be called a "4 solution." This is useful because, when we're dealing with something huge like a concentration of 1,000,000,000,000,000,000, it's easier to call it an "18 solution" rather than write the whole thing out.

So why does the scale run backward? The answer is that most acids, even very concentrated ones, only contain a small number of hydrogens per liter.

Most of the acids you're likely to encounter fall somewhere in the region of 0.00001 to 1. Writing this in standard form, we use negative powers, i.e. 10^{-6} to 10^{-1}. In this case, the 10^{-1} is the more concentrated solution.

It was a Danish chemist named Søren Sørensen who suggested we use negative logarithms when writing our acid concentrations, purely because it looks neater. The less concentrated acid gets a value of:

$$-\text{Log}_{10}\, 0.00001 = 6$$

While the more concentrated acid ends up as:

$$-\text{Log}_{10}\, 1 = 1$$

He referred to the negative logarithm of a number as the "potenz" of it, which really means "the power you have to raise the number ten to." And so we define the pH scale as:

$$\text{pH} = -\text{Log}_{10}\, (\text{concentration of hydrogen ions in a liter})$$

On this scale, most acids fall between 1 and 6 but the scale can go in either direction. An acid with a concentration of 10 would have a pH of 0, while an acid with a concentration of 100 would be pH −1.

The pK_a scale works in exactly the same way and uses the same system. Only this time, rather than measuring the concentration of hydrogens, we're measuring the strength of an acid, i.e. how willing it is to release a proton into solution. The best way to express this is to say what fraction of original acid ends up dissociating.

Suppose you have 100 acid molecules and only one of them splits. We would say the strength for this acid is 1 percent. For reasons that we won't get into (an appendix of an appendix is a bit silly), we express the strength of an acid as a fraction called the Ka. And, once again, these numbers are typically extremely small.

Only a tiny amount of hydrogens have the gumption to dissociate, so we get Ka values with high negative numbers. Using the p method, we take the negative logarithm of the Ka (how strong it is) and *voilà*, the result is our pK$_a$ scale.

Groups of the Periodic Table

Going from left to right across the periodic table, several of the columns (groups) have names that are largely historical. Groups 3 to 10 are named after the top element in the group e.g. group 10 is called "the nickel group," but all other columns have informal monikers.

Group 1 Alkali metals
Group 2 Alkaline earth metals
Group 3 Scandium group
Group 4 Titanium group
Group 5 Vanadium group
Group 6 Chromium group
Group 7 Manganese group
Group 8 Iron group
Group 9 Cobalt group
Group 10 Nickel group
Group 11 Coinage metals
Group 12 Volatile metals
Group 13 Icosagens
Group 14 Crystallogens
Group 15 Pnictogens
Group 16 Chalcogens
Group 17 Halogens
Group 18 Noble gases

Acknowledgments

One thing I've learned during the course of writing my first book is that while my name appears on the cover, what you read is a collaboration of many minds. I'd like to thank several of them.

First and foremost, the students of Northgate High School, for giving me a reason to get up in the morning and helping me get the whole thing off the ground. This book wouldn't exist without them. Rock and roll, guys. Rock and roll.

Huge and heartfelt thanks to my agent Jen Christie for taking a chance on an awkward writer, being patient when I was difficult, and for guiding me through the insane world of publishing and marketing—something far more complex than science.

Duncan Proudfoot from Little, Brown Book Group is wonderful and I want to thank him for immediately "getting" what I was trying to do. I really hope the book doesn't let him down.

In terms of the book's content, the person I need to thank most is Ella Catherall who edited every page, fact-checked the science, and corrected all 385 errors. If the book is any good, it's down to her.

Thank you to the Science department at Northgate (there's about twenty staff so I can't namecheck everyone) for all the tolerance they have shown me while I've learned to be a teacher. It's a privilege working with them. And in particular, a very special thanks to Hazel and David—not just for advice, but for their inspiration and friendship.

I need to thank the great Seishi Shimizu for teaching me how to think and write like a scientist. He was a great mentor and it was an honor being his student.

Thank you to Karl Dixon for reading and critiquing the manuscript, making me laugh when I stopped believing it was any good, and for being the Sherlock Holmes to my Watson these many, many years.

A huge thank you to Mandalyn King for always giving it to me straight and being pretty much amazing. I respect her opinion more than she knows.

From an earlier time in my life, I want to thank Mr. Evans for being *that* teacher, and John Miller for persuading me to become one myself.

I obviously need to thank my wife who is probably the most patient person in the world. Thanks for letting me spend so much time doing this.

And finally, I want to thank my father, who taught me the importance of asking questions and turned me into a scientist.

Notes

INTRODUCTION: A RECIPE FOR REALITY

1. R. W. Sterner, J. J. Elser, *Ecological Stoichiometry: The Biology of Elements from Molecules to the Biosphere* (Princeton, NJ: Princeton University Press, 2002).

CHAPTER 1: FLAME CHASERS

1. H. Krug, O. Ruff, "Uber ein neues chlorfuorid ClF$_3$," *Zeitschrift für anorganische und allgemeine Chemie*, vol. 190, no. 1 (1930), pp. 270–76.1.
2. "Compound summary for CID 24627," *Open Chemistry Database.* Available from: https://pubchem.ncbi.nlm.nih.gov/compound/chlorine_trifluoride#section=Top (accessed August 18, 2017).
3. J. D. Clark, *Ignition! An Informal History of Rocket Propellants* (New Brunswick, NJ: Rutgers University Press, 1972).
4. "Eastern Germany 2004," *Bunker Tours.* Available from: http://www.bunkertours.co.uk/germany_2004.htm (accessed August 18, 2017).
5. Diogenes Laertius, *The Lives and Opinions of Eminent Philosophers, Vol. II, Books 6–10,* trans. R. D. Hicks (Cambridge, MA: Harvard University Press, 1925).
6. "Protactinium," *Encyclopedia.* Available from: http://www.encyclopedia.com/science-and-technology/chemistry/compounds-and-elements/protactinium (accessed August 18, 2017).
7. J. Emsley, *The Shocking History of Phosphorus: A Biography of the Devil's Element* (London: Macmillan, 2000).
8. H. M. Leicester, H. S. Klickstein, *A Source Book in Chemistry 1400–1900* (Cambridge, MA: Harvard University Press, 1952).
9. H. Muir, *Eureka: Science's Greatest Thinkers and Their Key Breakthroughs* (London: Quercus, 2012).
10. Muir, *Eureka.*
11. M. Sędziwój, "Letters of Michael Sendivogius to the RoseyCrusian

Society," Epistle 54 (January 12, 1647), *The Masonic High Council the Mother High Council*. Available from: http://rgle.org.uk/Letters_Sendivogius.htm (accessed October 8, 2017).

12. I. Asimov, *Breakthroughs in Science* (Boston, MA: Houghton Mifflin, 1960).

13. R. Harré, *Great Scientific Experiments: Twenty Experiments that Changed Our View of the World* (Oxford: Phaidon, 1981).

14. I. Asimov, *Words of Science* (London: Harrap, 1974).

15. Isaiah 54:11.

16. "Periodic table—lithium," *Royal Society of Chemistry*. Available from: http://www.rsc.org/periodic-table/element/3/lithium (accessed August 18, 2017).

17. B. C. Gibb, "Hard-luck Scheele," *Nature Chemistry*, vol. 7 (2015), pp. 855–6.

18. Leicester and Klickstein, *A Source Book in Chemistry*.

CHAPTER 2: UNCUTTABLE

1. *The Core* (2003), dir. Jon Amiel, Paramount Pictures.

2. T. Irifune et al., "Ultrahard polycrystalline diamond from graphite," *Nature*, vol. 421 (2003), pp. 599–600.

3. David Robson, "How to make a diamond from scratch with peanut butter," *BBC* (November 7, 2014). Available from: http://www.bbc.com/future/story/20141106-the-man-who-makes-diamonds (accessed August 18, 2017).

4. B. Russell, *History of Western Philosophy* (Oxford: Routledge Classics, 2004).

5. D. Hurd, J. Kipling, *The Origins and Growth of Physical Science* (London: Penguin, 1958).

6. J. Dalton, *A New System of Chemical Philosophy* (London: R. Bickerstaff, 1808).

7. W. L. Masterson, C. N. Hurley, *Chemistry: Principles and Reactions* (Boston, MA: Cengage Learning, 2012).

8. R. Harré, *Great Scientific Experiments: Twenty Experiments that Changed Our View of the World* (Oxford: Phaidon, 1981).

9. A. Einstein, "Über die von der molekularkinetischen Theorie der Wärme geforderte Bewegung von in ruhenden Flüssigkeiten suspendierten Teilchen," *Annalen der Physik*, vol. 322 (1905), pp. 549–60.

CHAPTER 3: THE MACHINE GUN AND THE PUDDING

1. "A Boy and His Atom: The World's Smallest Movie," *IBM Research*. Available from: http://www.research.ibm.com/articles/madewith atoms.shtml (accessed August 18, 2017).
2. E. T. Whittaker, *A History of Theories of the Aether and Electricity* (Harlow: Longman, Green & Co, 1951).
3. E. Rutherford, *Nobel Lectures: Chemistry 1901–1921* (Amsterdam: Elsevier Publishing, 1966).
4. H. C. von Bayer, *Taming the Atom: The Emergence of the Visible Microworld* (New York: Random House, 1992).
5. R. W. Chabay, B. A. Sherwood, *Matter & Interactions*, third edition (Hoboken, NJ: Wiley, 2002).
6. *Man of Steel* (2013), dir. Zak Snyder, Warner Bros.
7. H. P. Lovecraft, *The Dunwich Horror and Other Stories* (London: Pocket Penguin Classics, 2010).
8. *Superman Returns* (2006), dir. Bryan Singer, Warner Bros; P. S. Whitfield et al., "$LiNaSiB_3O_7(OH)$—novel structure of the new borosilicate mineral jadarite determined from laboratory powder diffraction data," *Acta Crystallographica Section B*, vol. 63, no. 3 (2007), pp. 396–401.

CHAPTER 4: WHERE DO ATOMS COME FROM?

1. "The coldest place in the world," *NASA* (December 10, 2013). Available from: https://science.nasa.gov/science-news/science-at-nasa/2013/09 dec_coldspot (accessed August 18, 2017).
2. R. Sahai et al., "The coldest place in the Universe: Probing the ultra-cold outflow and dusty disk in the Boomerang Nebula," *The Astrophysical Journal*, vol. 841, no. 2 (2017).

3. J. W. Park et al., "Ultracold dipolar gas of fermionic Na23K40 molecules in their absolute ground state," *Physical Review Letters*, vol. 114 (2015).

4. Plato, *Theaetetus*, trans. J. McDowell (Oxford: Oxford University Press, 1999).

5. B. Russell, *History of Western Philosophy* (Oxford: Routledge Classics, 2004).

6. G. Dixon, P. Parsons, *The Periodic Table: A Field Guide to the Elements* (London: Quercus, 2013).

7. H. Aldersey-Williams, *Periodic Tales: The Curious Lives of the Elements* (London: Viking, 2011).

8. C. Payne-Gaposchkin, *Cecilia Payne-Gaposchkin: An Autobiography and Other Recollections* (Cambridge: Cambridge University Press, 1996).

9. "Cecilia Payne-Gaposchkin," *Encylopædia Britannica*. Available from: https://www.britannica.com/biography/Cecilia-Payne-Gaposchkin (accessed August 18, 2017).

10. "The early universe," CERN. Available from: https://home.cern/about/physics/early-universe (accessed August 18, 2017).

CHAPTER 5: BLOCK BY BLOCK

1. "The Scoville Unit," *Jalapeño Madness*. Available from: http://www.jalapenomadness.com/jalapeno_scoville_units.html (accessed August 18, 2017).

2. "'World's hottest' chilli pepper grown in St Asaph," *BBC News* (May 17, 2017). Available from: http://www.bbc.com/news/uk-wales-north-east-wales-39946962 (accessed August 18, 2017).

3. A. Szallasi, P. M. Blumberg, "Resiniferatoxin, a phorbol-related diterpene, acts as an ultrapotent analog of capsaicin, the irritant constituent in red pepper," *Neuroscience*, vol. 30, no. 2 (1989), pp. 515–20.

4. "How we taste," *Technology Review* (April 2004). Available from: https://www.heise.de/tr/artikel/Wie-wir-schmecken-404206.html (accessed August 18, 2017).

5. "Vantablack," *Surrey Nanosystems*. Available from: https://www.surreynanosystems.com/vantablack (accessed August 18, 2017).

6. J. Clayden, N. Greeves, S. Warren, *Organic Chemistry*, second edition (Oxford: Oxford University Press, 2012); "4 workers killed at DuPont Chemical plant," *Scientific American* (November 18, 2014). Available from: https://www.scientificamerican.com/article/4-7.workers-killed -at-dupont-chemical-plant (accessed August 18, 2017).

7. B. Russell, *History of Western Philosophy* (Oxford: Routledge Classics, 2004).

8. B. Pennington, "The death of Pythagoras," *Philosophy Now*, no. 121 (2017).

9. Russell, *History of Western Philosophy*.

10. A. Lavoisier, *Traite Elementaire de Chemie* (Paris: Cuchet, 1789).

11. E. Scerri, *The Periodic Table: Its Story and Its Significance* (Oxford: Oxford University Press, 2006).

12. E. Scerri, *The Periodic Table: A Very Short Introduction* (Oxford: Oxford University Press, 2011).

13. J. E. Jorpes, *Jac. Berzelius: His Life and Work* (Stockholm: Royal Swedish Academy of Science, 1966).

14. J. A. R. Newlands, *On the Discovery of the Periodic Law: and On Relations of the Atomic Weights* (London: E. & F. N. Spon, 1884).

15. M. D. Gordin, *A Well-Ordered Thing: Dmitrii Mendeleev and the Shadow of the Periodic Table* (New York: Basic Books, 2004).

16. "Periodic Law," *Mendeleev*. Available from: http://www.mendeleev. nw.ru/period_law/ver_trif.html (accessed August 18, 2017).

CHAPTER 6: QUANTUM MECHANICS SAVES THE DAY

1. A. Werner, "Beitrag zum Ausbau des periodischen systems," *Berichte der deutschen chemischen Geselkchaft*, vol. 38 (1905), pp. 914–21.

2. G. Seaborg, "Priestley Medal Address—The Periodic Table: Tortuous Path to Man-Made Elements" (April 16, 1979), reprinted in G. Seaborg, *Modern Alchemy: Selected Papers of Glenn Seaborg Vol. 2* (Singapore: World Scientific Publishing Co., 1994).

3. H. E. White, *Introduction to Atomic Spectra* (New York: McGraw-Hill, 1934).

4. E. H. Riesenfeld, *Practical Inorganic Chemistry*, reprint of the 1943 edition (Barcelona: Labour, 1950).
5. Seaborg, "Priestley Medal Address."

CHAPTER 7: THINGS THAT GO BOOM

1. T. M. Klapötke et al., "New azidotetrazoles: Structurally interesting and extremely sensitive," *Chemistry—An Asian Journal*, vol. 7, no. 1 (2012), pp. 214–24.
2. "Alfred Nobel," *Encylopædia Britannica*. Available from: https://www.britannica.com/biography/Alfred-Nobel (accessed August 18, 2017); E. J. Sirleaf, "Alfred Nobel's legacy to women," *New York Times* (December 12, 2011).
3. "Alfred Nobel's fortune," *Nobel Peace Prize*. Available from: https://www.nobelpeaceprize.org/History/Alfred-Nobel-s-fortune (accessed August 18, 2017).
4. J. Janes, *Documents which Changed the Way We Live* (Lanham, MD: Rowman & Littlefield, 2017).
5. K. Fant, *Alfred Nobel: A Biography* (New York: Arcade Publishing, 2014).

CHAPTER 8: THE ALCHEMIST'S DREAM

1. "Sotheby's sells record $71 million diamond to Chow Tai Fook," *Bloomberg* (April 4, 2017). Available from: https://www.bloomberg.com/news/articles/2017-04-04/sotheby-s-sets-world-record-selling-71-million-pink-diamond (accessed August 18, 2017).
2. R. Kurin, *Hope Diamond: The Legendary History of a Cursed Gem* (New York: HarperCollins, 2007).
3. "Plutonium certified reference materials price list," *US Department of Energy—Office of Science*. Available from: https://science.energy.gov/nbl/certified-reference-materials/prices-and-certificates/plutonium-certified-reference-materials-price-list (accessed August 18, 2017).

4. "Californium price," *Metalary*. Available from: https://www.meta-lary.com/californium-price (accessed August 18, 2017).

5. G. D. Hedesan, *An Alchemical Quest for Universal Knowledge: The "Christian Philosophy" of Jan Baptist Van Helmont 1579–1644* (Oxford: Routledge, 2016).

6. R. Patai, *The Jewish Alchemists: A History and Source Book* (Princeton, NJ: Princeton University Press, 1994).

7. B. Jonson, *The Alchemist* (1610). Available from: http://www.public-library.uk/ebooks/14/35.pdf (accessed August 18, 2017).

8. S. Lee, S. Ditko, *Amazing Fantasy*, no. 15 (August 15, 1962); S. Lee, J. Kirby, *The Incredible Hulk*, no. 1 (May 1, 1962); S. Lee, J. Kirby, *The Fantastic Four*, no. 1 (November 1, 1961); S. Lee, B. Everett, *Daredevil*, no. 1 (April 1, 1964); C. Claremont, J. Byrne, *X-Men*, no. 137 (September 1, 1980), and *Phoenix: The Untold Story* (April 1, 1984).

9. *Godzilla* (1954), dir. Ishiro Honda, Toho Co. Ltd.

10. C. Patterson, "Age of meteorites and the earth," *Geochimica et Cosmochimica Acta*, vol. 10, no. 4 (1956), pp. 230–7.

11. E. Rutherford, "The Collision of Alpha-particles with Light Atoms," *Philosophical Magazine*, vol. 37 (1919).

12. "Public ignorant about radiation dose of mammograph," *Medscape* (May 12, 2014). Available from: http://www.medscape.com/viewarticle/824999 (accessed August 18, 2017).

13. Gary Mansfield, "Banana equivalent dose" (March 7, 1995). Available from: http://health.phys.iit.edu/extended_archive/9503/msg00074.html (accessed August 18, 2017).

14. D. R. Corson, K. R. MacKenzie, E. Serge, "Artificially radioactive element 85," *Physical Review*, vol. 58, no. 8 (1940), pp. 672–8.

15. *Iron Man 2* (2010), dir. Jon Favreau, Paramount Pictures.

16. "Edwin M. McMillan—facts," *Nobel Prize*. Available from: http://www.nobelprize.org/nobel_prizes/chemistry/laureates/1951/mcmillan-facts.html (accessed August 18, 2017).

17. R. M. Shoch, *Case Studies in Environmental Science* (Eagan, MN: West Publishing Co., 1996).

18. "Americium," *ACS Publications*. Available from: http://pubs.acs.org/cen/80th/print/americiumprint.html (accessed August 18, 2017).

19. "IUPAC announces the names of the elements 113, 115, 117 and 118," *International Union of Pure and Applied Chemistry* (November 30, 2016. Available from: https://iupac.org/iupac-announces-the-names-of-the-elements-113-115-117-and-118 (accessed August 18, 2017).

20. J. Emsley, *Nature's Building Blocks: An A–Z Guide to the Elements* (Oxford: Oxford University Press, 2001).

CHAPTER 9: LEFTISTS

1. A. K. Geim, M. V. Berry, "Of flying frogs and levitrons," *European Journal of Physics*, vol. 18, no. 4 (1997), pp. 307–13.

2. K. S. Novoselov et al., "Electric firled effect in atomically thin carbon films," *Science*, vol. 306, no. 5696 (2004), pp. 666–9.

3. "How strong is graphene?," *University of Manchester*. Available from: http://www.graphene.manchester.ac.uk/discover/video-gallery/what-is-graphene/how-strong-is-graphene (accessed August 18, 2017); J. Abraham et al., "Tunable sieving of ions using graphene oxide membranes," *Nature Nanotechnology*, no. 12 (2017), pp. 546–50.

4. "Properties of stainless steel, metals and other conductive materials," *TibTech Innovations*. Available from: http://www.tibtech.com/conductivity.php (accessed August 18, 2017); "Understanding graphene," *Graphenea*. Available from: https://www.graphenea.com/pages/graphene (accessed August 18, 2017).

5. J. Romer, *A History of Ancient Egypt: From the First Farmers to the Great Pyramid* (New York: Thomas Dunne Books, 2013).

6. J. Levy, *Scientific Feuds: From Galileo to the Human Genome Project* (London: New Holland Publishers, 2010).

7. S. Gray, "An account of some new electrical experiments," *Philosophical Transactions of the Royal Society of London*, vols 31–3 (1708).

8. D. S. Lemons, *Drawing Physics: 2,600 Years of Discovery from Thales to Higgs* (Cambridge, MA: MIT Press, 2017).

9. P. Bertucci, "Sparks in the dark: The attraction of electricity in the eighteenth century," *Endeavour*, vol. 31, no. 3 (2007).

10. C. Brandon, *The Electric Chair: An Unnatural American History* (Jefferson, NC: McFarland, 1999).

11. Levy, *Scientific Feuds*; *Electrocuting an Elephant (1903)—WARNING: Viewer Discretion—Disturbing footage—Thomas Edison*, Change Before Going Productions (January 16, 2014). Available from: https://www.youtube.com/watch?v=NoKi4coyFw0 (accessed August 18, 2017).

12. C. S. Combs, *Deathwatch: American Film, Technology and the End of Life* (New York: Columbia University Press, 2014).

13. M. S. Rosenwald, "'Great God, he is alive!' The first man executed by electric chair died slower than Thomas Edison expected," *Washington Post* (April 28, 2017).

CHAPTER 10: ACIDS, CRYSTALS, AND LIGHT

1. D. Wilson, *A History of British Serial Killing* (London: Sphere, 2011); M. Whittington-Egan, R. Whittington-Egan, *Murder on File: The World's Most Notorious Killers* (Castle Douglas: Neil Wilson Publishing, 2005).

2. D. H. Ripin, D. A. Evans, "pK_as of inorganic and oxo-acids," *The Evans Group*. Available from: http://evans.rc.fas.harvard.edu/pdf/evans_pKa_table.pdf (accessed August 18, 2017).

3. Ripin, Evans, "pK_as of inorganic and oxo-acids"; G. T. Cheek, "Electrochemical studies of the Fries rearrangement in ionic liquids," *Electrochemical Society Transactions*, vol. 16, no. 49 (2009), pp. 541–4.

4. G. A. Olah, "My search for carbocatins and their role in chemistry," Nobel Lecture (December 8, 1994).

5. To illustrate this point, the author has taken the claim from the article on superacids from *Wikipedia*. Available from: https://en.wikipedia.org/wiki/Superacid (accessed August 18, 2017) to illustrate this point. *Wikipedia* cites G. A. Olah, "Crossing conventional boundaries in half a century of research," *Journal of Organic Chemistry*, vol. 70, no. 7 (2005), pp. 2413–29, for the claim that fluoroantimonic acid is 10^{16} times stronger than sulfuric known to have a pK_a of –3, giving a pK_a of –19.

6. T. R. Hogness, E. G. Lunn, "The ionisation of hydrogen by electron impact as interpreted by positive ray analysis," *Physical Review*, vol. 21, no. 1 (1925), pp. 44–55.

7. This number is calculated from generating a Born-Haber cycle via: S. Lias et al., "Evaluated gas phase basicities and proton affinities of molecules: Heats of formation of protonated molecules," *Journal of Physical and Chemical Reference Data*, vol. 13, no. 3 (1984), p. 695, and assumes that the HHe$^+$ ion has a similar solubility to a lithium ion, which has comparable size. If we assume a free energy change of dissociation to be −360 kJmol^{-1} then at standard temperature and pressure we can invoke G = −RT lnKa. Taking −360/(0.008314 × 273) we obtain 158.6 = lnKa and therefore a Ka to have value of 4.15 x10^{68}. Taking the negative logarithm of this number yields −68.6, which the author has rounded to −69.

8. "Strange but true: Superfluid helium can climb walls," *Scientific American* (February 20, 2009). Available from: https://www.scientificamerican.com/article/superfluid-can-climb-walls (accessed August 18, 2017).

CHAPTER 11: IT'S ALIVE, IT'S ALIVE!

1. A. C. Nathwani et al., "Polonium-210 poisoning: a first-hand account," *The Lancet*, vol. 388, no. 10049 (2016), pp. 1075–80.

2. R. H. Adamson, "The acute lethal dose 50 (LD50) of caffeine in albino rats," *Regulatory Toxicology and Pharmacology*, vol. 80 (2016), pp. 274–6.

3. E. Welsome, *The Plutonium Files: America's Secret Medical Experiments in the Cold War* (New York: The Dial Press, 1999).

4. Lead: K. Sujatha et al., "Lead acetate induced neurotoxicity in Wistar albino rats: A pathological, immunological, and ultrastructural study," *Journal of Pharma and Bio Science*, no. 2 (2011), pp. 459–62. Note: this assumes lead acetate. Thallium: Agency for Toxic Substances and Disease Registry, *Toxicological Profile for Thallium* (Atlanta, GA: Agency for Toxic Substances and Disease Registry, 1992). Available from: https://www.atsdr.cdc.gov/ToxProfiles/tp.asp?id=309&tid=49 (accessed August 18, 2017). Note: this assumes thallium acetate for fair comparison with lead. Arsenic: H. Marquardt et al., *Toxicology* (Cambridge, MA: Academic Press, 1999). Phosphorus: Agency for

Toxic Substances and Disease Registry, *Toxicological Profile for White Phosphorus* (Atlanta, GA: Agency for Toxic Substances and Disease Registry, 1997) Available from: https://www.atsdr.cdc.gov/toxprofiles/tp103-c2.pdf (accessed August 18, 2017). Note: the value quoted seems to come from C. C. Lee, *Mammalian Toxicity of Munition compounds. Phase I: Acute Oral Toxicity, Primary Skin and Eye Irritation, Dermal Sensitization, and Disposition and Metabolism*, Report No. 1, AD B011150 (Kansas City, MO: Midwest Research Institute, 1975).

5. S. Ela, "Experimental study of toxic properties of dimethylcadmium," *Gigiena Truda i Professional'nye Zabolevaniya*, no. 6 (1991), pp. 14–17.

6. J. R. Barash, S. S. Arnon, "A novel strain of clostridium botulinum that produces Type B and Type H botulinum toxins," *The Journal of Infectious Diseases*, vol. 29, no. 2 (2014), pp. 183–91.

7. "Botox OnabotuliniumtoxinA," *Botox*. Available from: http://www.botox.com (accessed August 18, 2017).

8. C. H. Mayo, interview given in *Northwestern Health Journal* (December 1924).

9. V. Busacchi, "Vincenzo Menghini and the discovery of iron in the blood," *Bullettino delle science mediche*, vol. 130, no. 2 (1958), pp. 202–5.

10. E. Kinne-Saffran, R. K. Kinne, "Vitalism and synthesis of urea. From Friedrich Wöhler to Hans A. Krebs," *American Journal of Nephrology*, vol. 19, no. 2 (1999), pp. 290–4.

11. K. H. Antman, "Introduction: The history of arsenic trioxide in cancer therapy," *The Oncologist*, vol. 6, no. 2 (2001), pp. 1–2.

12. N. C. Lloyd, "The composition of Ehrlich's salvarsan: Resolution of a century-old debate', *Angewandte Chemie*, vol. 44, no. 6 (2005), pp. 941–4.

13. H. P. Chauhan, "Synthesis, spectroscopic characterization and antibacterial activity of antimony(III)bis(dialkyldithiocarbamato) alkyldithiocarbonates," *Spectrochimica Acta. Part A*, vol. 81, no. 1 (2011), pp. 417–23; "Education in Chemistry—Cerium," *Royal Society of Chemistry*. Available from: https://eic.rsc.org/elements/cerium/2020005.article (accessed August 18, 2017).

14. "Getting a tiny bit of this element on your skin will make you reek

of garlic for weeks," *io9* (August 13, 2015). Available from: http://io9.gizmodo.com/getting-a-tiny-bit-of-this-element-on-your-skin-will-ma-1723949124 (accessed August 18, 2017).

15. R. Hambrecht et al., "Managing your angina symptoms with nitroglycerin," *Circulation*, no. 127 (2013).

16. V. S. Ramachandran, *Encyclopedia of the Human Brain* (Cambridge, MA: Academic Press, 2002).

17. T. Bartholin, *Historiarum anatomicarum rariorum centuria I et II* (1654). Available from: https://books.google.nl/books?id=NT-LAd44hZ4UC&printsec=frontcover&dq=%22Historiarum+anatomicarum+rariorum+centuria+I%22&hl=en&sa=X&ei=6T-MLVagK09SgBJvGgaAH&redir_esc=y#v=onepage&q=%22Historiarum%20anatomicarum%20rariorum%20centuria%20I%22&f=false (accessed August 18, 2017).

18. "New light on human torch mystery," *BBC News* (31 August 1998). Available from: http://news.bbc.co.uk/2/hi/uk_news/158853.stm (accessed August 18, 2017).

19. M. Harrison, *Fire from Heaven: A Study of Spontaneous Combustion in Human Beings* (London: Skoob Books, 1990).

20. "Cause of fire killing woman still mystery," *St. Petersburg Times*, Section 2 (July 4, 1951). Available from: https://news.google.com/newspapers?nid=888&dat=19510704&id=rwRZAAAAIBAJ&sjid=lE8DAAAAIBAJ&pg=3085,1265930&hl=en (accessed August 18, 2017).

21. Garth Haslam, "1951, July 1: Mary Reeser's fiery death," *Anomalies: The Strange and Unexplained*. Available from: http://anomalyinfo.com/Stories/1951-july-1-mary-reesers-strange-death (accessed August 18, 2017).

22. L. E. Arnold, *Ablaze! The Mysterious Fires of Spontaneous Human Combustion* (New York: M. Evans and Co., 1995).

23. J. Randles, P. Hough, *Spontaneous Human Combustion* (London: Robert Hale Ltd, 2007).

24. G. Whitley, "Garston Church" (1867–74), *Speke Archive Online*. Available from: http://spekearchiveonline.co.uk/garston_church.htm (accessed August 18, 2017).

25. G. Gassmann, D. Glindemann, "Phosphane (PH_3) in the biosphere," *Angewandte Chemie*, vol. 32, no. 5 (1993), pp. 761–3.

CHAPTER 12: NINE ELEMENTS THAT CHANGED THE WORLD (AND ONE THAT DIDN'T)

1. "Pitch Drop Demonstration," *National Museums Scotland.* Available from: https://www.nms.ac.uk/explore-our-collections/stories/science-and-technology/made-in-scotland-changing-the-world/scottish-science-innovations/pitch-drop-demonstration (accessed September 9, 2017).
2. "Bart the Lover," *The Simpsons*, season 3, episode 16, dir. Carlos Baeza (original airdate February 13, 1992).
3. J. Emsley, *Nature's Building Blocks: An A–Z Guide to the Elements* (Oxford: Oxford University Press, 2001).
4. E. Barrett, J. Mingo, *Not Another Apple for the Teacher: Hundreds of Fascinating Facts from the World of Education* (Newburyport, MA: Conari Press, 2002).
5. "The story of how the tin can nearly wasn't," *BBC News* (April 21, 2013). Available from: http://www.bbc.com/news/magazine-21689069 (accessed August 18, 2017).
6. Adapted from "Gold fun facts," *American Museum of Natural History.* Available from: http://www.amnh.org/exhibitions/gold/eureka/gold-fun-facts (accessed August 18, 2017).
7. Adapted from R. O'Connell et al., *GFMS Gold Survey 2016* (New York: Thomson Reuters, 2016).
8. "The history of money," *The Mint of Finland.* Available from: https://www.suomenrahapaja.fi/eng/about_money/the_history_of_money (accessed August 18, 2017); E. M. Green, *Lady Midrash: Poems Reclaiming the Voices of Biblical Women* (Eugene, OR: Wipf and Stock, 2016).
9. J. O. Nriagu, "Saturnine gout among Roman aristocrats—did lead poisoning contribute to the fall of the empire?," *New England Journal of Medicine*, no. 308 (1983), pp. 660–3.
10. H. Needleman, "Low level lead exposure: History and discovery,"

Annals of Epidemiology, vol. 19, no. 4 (2009), pp. 235–8; H. Delile et al., "Lead in ancient Rome's city waters," *PNAS*, vol. 11, no. 18 (2014), pp. 6594–9.

11. D. Childress, *Johannes Gutenberg and the Printing Press* (Minneapolis, MN: Twenty First Century Books, 2008).

12. A. Gallop, "Mortality improvements and evolution of life expectancies," *Actuary, Pensions Policy, Demography and Statistics* (London: Government Actuary's Department, 2006).

13. G. W. Beardsley, "The 1832 cholera epidemic," *Early America Review*, vol. 4, no. 1 (2000).

14. "Measles" and "Frequently asked questions and answers on smallpox," *World Health Organization*. Available from: http://www.who.int/mediacentre/factsheets/fs286/en/ and available from: http://www.who.int/csr/disease/smallpox/faq/en (accessed August 18, 2017).

15. D. Charles, *Between Genius and Genocide: The Tragedy of Fritz Haber, Father of Chemical Warfare* (London: Jonathan Cape, 2005).

16. "How much water does the average person use at home per day?," *United States Geological Survey*. Available from: https://water.usgs.gov/edu/qa-home-percapita.html (accessed August 18, 2017).

17. T. P. Garrett, "The wonderful development of photography," *The Art World*, vol. 2, no. 5 (1917), pp. 489–91.

18. Stephen Herbert, "Wordsworth Donisthorpe," *Who's Who of Victorian Cinema* (2000). Available from: http://www.victorian-cinema.net/donisthorpe (accessed August 18, 2017).

19. J. Watson, *DNA: The Secret of Life* (London: Arrow Books, 2003).

20. "U.S. Nuclear Weapons Capability," *2017 Index of U.S. Military Strength* (2017). Available from: http://index.heritage.org/military/2017/assessments/us-military-power/u-s-nuclear-weapons-capability (accessed August 18, 2017).

21. *J. Robert Oppenheimer: "I am become death, the destroyer of worlds,"* Plenilune pictures, (August 6, 2011). Available from: https://www.youtube.com/watch?v=lb13ynu3Iac (accessed August 18, 2017).

22. J. N. Shurkin, *Broken Genius: The Rise and Fall of William Shockley, Creator of the Electronic Age* (London: Macmillan, 2006).

23. A. Usanov et al., *Coltan, Congo & Conflict: Polinares Case Study* (The Hague: The Hague Centre for Strategic Studies, no. 21.05.13, 2013); E. Sutherland, "Coltan, the Congo and your cell phone: the connection between your mobile phone and human rights abuses in Africa," *MIT* (2016). Available from: http://web.mit.edu/12.000/www/m2016/pdf/coltan.pdf (accessed August 18, 2017).

24. D. Grossmann, C. Ganz, P. Russell, *Zeppelin Hindenburg: An Illustrated History of LZ-129* (Stroud: The History Press, 2017).

25. O. A. Hurricane et al., "Fuel gain exceeding unity in an inertially confined fusion implosion," *Nature*, vol. 506 (2014), pp. 343–7.

Index